TURBINE-ENGINED AIRLINERS OF THE WORLD

F. G. SWANBOROUGH

 TEMPLE PRESS BOOKS · LONDON 1962

First published 1962

© TEMPLE PRESS BOOKS LIMITED, 1962

MADE AND PRINTED IN GREAT BRITAIN
FOR
TEMPLE PRESS BOOKS LIMITED
BOWLING GREEN LANE, LONDON EC1
BY
Morrison & Gibb Ltd.
London and Edinburgh

CONTENTS

INTRODUCTION

MORE THAN 112 million passengers were carried by the world's airlines in 1961, compared with 9 million in 1945. In the fifteen intervening years, the airline industry had expanded into a major world force, providing fast communications from country to country and making a unique contribution to national economy.

For a decade since 1952, the airlines have been introducing a new generation of airliners powered by gas turbine engines. This group includes both turboprop and turbojet types, the former having propellers turned by the gas turbine, the latter being the pure-jet types. Each has played its part in the air transport revolution which is still continuing.

This volume contains information on all the turboprop and turbojet airliners which have gone into regular service with the airlines up to the middle of 1962, as well as some newer types now in production. The text describes the development history of each design and records significant dates and airline orders, and full technical details are presented in a standard form for easy reference and direct comparison.

The illustrations have been chosen both for their pictorial quality and also to show the markings of the different airlines which use or have used each type. Details of the production of each type are contained in a separate section.

Most of the 300 photographs in this volume have been provided by the manufacturers of the aircraft or by the airlines, to whose public relations staffs thanks are due. The author also acknowledges gratefully the assistance of Peter Keating, whose pictures filled a number of gaps, and of J. M. G. Gradidge and Peter March, who contributed valuable information to the Production Record.

F. G. SWANBOROUGH

ANTONOV An-24 (U.S.S.R.)

EXPECTED TO BE ready for service with the Russian state airline Aeroflot before the end of 1962, the Antonov An-24 is a product of the third stage of Soviet post-war re-equipment. The first stage of this programme was represented by the development of the turbojet Tupolev Tu-104 (q.v.), followed by the introduction of the large turboprop types, the Ilyushin Il-18 (q.v.) and Antonov An-10. The third stage called for modern short-haul equipment to take over from the piston-engined Il-12s and -14s on domestic routes throughout Russia, and therefore capable of operating from small airfields with no runways.

To meet this requirement the design bureau headed by Oleg K. Antonov produced an aeroplane which is essentially a scaled-down version of the An-10, and is in the same category as the Handley Page Herald and Fokker F-27. Its existence became known outside Russia in 1960 when photographs of the prototype (CCCP-L1959) appeared. The first flight was made by G. I. Lysenko and Yu. V. Kurlin, probably at the end of 1959 or in the early months of 1960.

The An-24 was seen from the photographs to follow conventional Antonov lines, with a high wing having anhedral on its outer panels and a large tail unit with a dorsal fin and dihedral tailplane. The engines are reported to be 2,500 h.p. Ivchenko AI-24 turboprops, turning four-blade propellers.

In the course of subsequent flight trials with several examples of the An-24, including a second prototype (CCCP-L1960), the size of the dorsal fin was increased and a small ventral fin was added. The engine nacelles were lengthened to project aft of the wing trailing edge and the fuselage nose also was lengthened slightly, perhaps to accommodate weather radar.

The basic layout in the An-24's pressurized fuselage is for thirty-two passengers in eight rows of double seats matching the eight windows. Accommodation can be increased to forty-four by reducing the seat pitch. The crew normally comprises two pilots, a navigator and a cabin attendant. A galley and baggage hold are usually located between the flight deck and the main cabin, in the plane of the propeller discs.

Dimensions
Span: 95 ft. 9½ in.
Overall length: 77 ft. 2½ in.
Overall height: 27 ft. 3½ in.
Gross wing area: 767 sq. ft.
Internal cabin dimensions:
    Length (ex. flight deck): 48 ft. 1½ in.
    Max. width: 111 in.; max. height: 77 in.
Max. usable floor area: 430 sq. ft.
Max. usable volume: 2,750 sq. ft.

Accommodation
Normal standard: 44 (4 abreast at 31-in. pitch).

Powerplants
Two Ivchenko AI-24 turboprops.
Take-off power each (I.S.A., s.l.): 2,500 e.h.p.

Weights and loadings
Total fuel: 9,880 lb.
Mfrs. max. payload: 8,820 lb.
Max. take-off: 42,330 lb.
Wing loading (max. t-o. wt.): 56.2 lb./sq. ft.
Power loading (max. t-o. wt.): 8.47 lb./e.h.p.

Performance
Continuous cruising speed (I.S.A.): 256 knots.
Take-off run: 1,500 ft.
Landing run: 1,500 ft.
Max. range (1,220 imp. gal.): 1,100 n. mi.

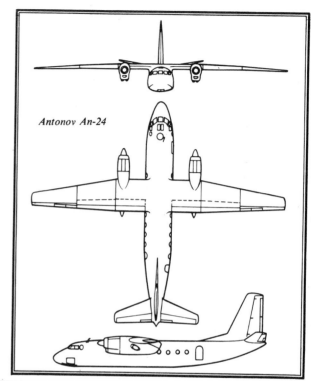

Antonov An-24

An early picture of the prototype Antonov An-24, before the addition of the ventral fin, showing the original short nacelles.

CCCP-Л1959

ANTONOV An-10 (U.S.S.R.)

An Aeroflot Antonov An-10A photographed over Moscow in 1961.

SINCE IT WENT into regular service with Aeroflot on July 22, 1959, the An-10 has made an important contribution to the growth of air traffic in Russia. It was designed by the bureau headed by Oleg K. Antonov to a stringent requirement which combined military and civil needs for a heavy freighter capable of operating from grass airfields of comparatively small size.

Work on the design began in November, 1955, after the bureau had completed the An-8 twin-turboprop transport for the Soviet Air Force, and the general lines of the An-8 were retained. The prototype was rolled out of Antonov's principal construction plant in Siberia at the end of 1956 and began its flight trials early the following year. When the first photographs were released it was seen to have a high wing on a circular-section fuselage which is believed to have the largest cross section of any pressurized cabin to date. The bogie main undercarriage units with low-pressure tyres for rough-field operation retracted into fuselage-side fairings which also housed the cabin pressurizing equipment and much of the hydraulic system, after the manner of the Douglas C-133.

Accommodation was provided for eighty-four passengers in three cabins in the first examples of the An-10 but this was increased to one hundred in the basic production version. With a diameter of 13½ ft. the fuselage has ample width for six-abreast seating and provides a headroom of 9 ft. in the cabin.

Before the An-10 entered service it underwent more than two years of flight development, in the course of which several important modifications were made to improve the lateral control. These modifications included the addition of small end-plate fins on the tailplane, and re-jigging of the wing to have anhedral on the outer panels. In this development programme also, the relative merits of the Ivchenko AI-20 and Kuznetsov NK-4 turboprops were assessed in flight and the former were adopted as the standard in the production aircraft. Small external differences were visible on the nacelles for the two types of engine, which were of similar power.

Before going into scheduled passenger service, the An-10 underwent a prolonged proving programme carrying freight on routes in Siberia. The first passenger flights, starting on July 22, 1959, were on the routes linking Simferopol with Moscow and Kiev.

In February, 1960, the 100-seat An-10A entered service. This variant incorporated the various improvements found necessary in the course of the proving flights, and is also reported to have a two-foot fuselage extension. A further development, with a fuselage lengthened by 10 ft. to provide a basic layout for 132 passengers, carries the design bureau designation An-16 but is expected to go into service as the An-10B.

During 1961, with the An-10A firmly established in service within Russia, particularly on routes in the more remote areas, the type was supplied to the Polar Aviation Division of Aeroflot and began operating regular supply missions to polar bases. For these operations, massive skis have been developed for the An-10A, which is one of the largest and heaviest aircraft ever to operate on skis.

In December, 1961, an An-10A (CCCP-34366), in company with an Il-18, made a pioneering flight from Russia through India, Indonesia, Australia and New Zealand to the Soviet Antarctic base at Mirny, near the South Pole, and regular flights were expected to follow over this route also. Another event of 1961 was the setting of a record speed in the FAI category for propeller-driven aircraft over a 500-km. circuit, at 454 m.p.h.

The military version of the An-10 is designated An-12 and is distinguished principally by having a rear-loading ramp in the underside of the fuselage, and a gun turret in the fuselage under the rudder. This version was in service with the Soviet Air Force by 1960 and some examples were seen outside Russia

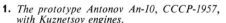

1. *The prototype Antonov An-10, CCCP-1957, with Kuznetsov engines.*
2. *A pre-production An-10, CCCP-11158, with anhedral on the outer wings, end-plate fins and a larger dorsal fin.*
3. *An Aeroflot An-10A photographed on its first visit to the U.K.*
4. *Ghana Airways' An-12 9G-AAZ, complete with rear turret position.*

5

carrying civil marking. In 1961, delivery began of sixteen examples ordered by the Indian Air Force.

At least one example of the An-12 has also been delivered to Ghana and this aircraft, 9G-AAZ, used by Ghana Airways as a

---

5. *A military An-12 with rear turret, carrying a civil registration for an overseas flight.*

---

freighter, is the only known example of the Antonov design in foreign airline service. This aircraft is illustrated on this page.

The number of An-10s built for Aeroflot is uncertain. One sequence of registration numbers appears to run between CCCP-11135 and CCCP-11229 (ninety-five aircraft if the numbers are consecutive). A later batch includes CCCP-34366 and CCCP-34385, so more than one hundred may have been built by the end of 1961, in addition to military versions.

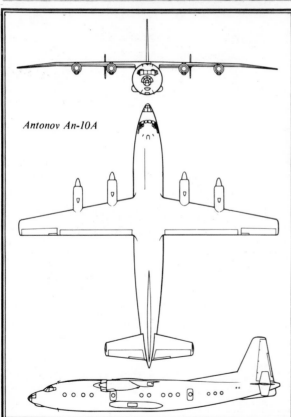

Antonov An-10A

ANTONOV **An-10**

Dimensions

Span: 124 ft. 8 in.
Overall length: 121 ft. 4 in.
Overall height: 32 ft. 1 in.
Gross wing area: 1,293 sq. ft.
Sweepback: Nil.
Internal cabin dimensions:
  Max. width: 162 in.; max. height: 108 in.

Accommodation

Normal tourist: 100 (5 abreast).
Max. coach: 126 (6 abreast).
Volume of freight and baggage holds: 1,483 cu. ft.

Powerplants

Four Ivchenko AI-20 turboprops.
Take-off power each (I.S.A., s.l.): 4,015 e.h.p.
Four-blade 14-ft. 9-in. propellers, reversible pitch.

Weights and loadings

Total fuel: 22,600 lb.
Mfrs. max. payload: 32,000 lb.
Max. take-off: 121,500 lb.
Wing loading (max. t-o. wt.): 88.0 lb./sq. ft.

Performance

Continuous cruising speed (I.S.A.): 325 knots at 26,000 ft.
Take-off run (at max. t-o. wt., I.S.A., s.l.): 2,300 ft.
Landing run (max. landing wt.): 1,640 ft.
Range (take-off to landing), still air, 1-hr. fuel reserves, I.S.A., max. fuel (2,800 imp. gal.): 1,850 n. mi. with 18,600-lb. payload.
Range (take-off to landing), still air, 1-hr. fuel reserve, I.S.A., with max. payload, 1,080 n. mi.

Structure. Circular-section semi-monocoque stressed-skin fuselage. Five-section two-spar box wing, with centre section continuous through fuselage.

Fuel system. Bag tanks in wing centre section, with gravity refuelling. Capacity approx. 2,800 imp. gal.

Undercarriage. Four-wheel main bogies with 1,050 × 300-mm. tyres retract upwards into fuselage fairings. Hydraulic disc brakes. Steerable (35 deg. each way) twin nosewheel. Retractable tail skid.

Flying controls. Manual system with aerodynamically balanced ailerons, elevators and rudder. Electric trimmers.

Flaps, air brakes. Fowler-type double-slotted flaps, max. deflection 41 deg., in two sections each side. No air brakes.

Cabin conditioning. Pressurization at 7.1 p.s.i. by engine-bleed air. System components in wheel fairings.

Hydraulics. Duplicated 2,140-p.s.i. systems, each powered by two engine-driven pumps, for u/c, flaps, nosewheel steer, brakes and windscreen wipers.

De-icing. Electrothermal de-icing of wing, tailplane, fin, propellers, intakes and windscreen.

ARMSTRONG WHITWORTH **ARGOSY** (Great Britain)

*The Armstrong Whitworth Argosy G-APRN was the fourth to fly, and was later converted for use by BEA as **an** Argosy 102.*

WHEN RIDDLE AIRLINES, an American air-freighting specialist, put the Armstrong Whitworth Argosy into regular service at the end of 1960, it became the world's first turbine-engined pure freighter in operation. The event marked the culmination of nearly four years of privately financed development of the aircraft in a programme which cost the Hawker Siddeley Group £13 million.

The decision to proceed with the AW650 design for a civil freighter was taken during 1956, after project work had been completed on the AW66 as a military freighter. The AW650 retained a similar layout to the AW66, with a pressurized fuselage having loading doors at the front and rear, and twin booms carrying the tailplane. The flight deck was located above the nose door to leave the main cabin space uninterrupted.

To save both time and expense, the original AW650 used an adaptation of the Avro Shackleton wing, redesigned by Avro as the Type 733 to mount four Rolls-Royce Dart turboprops in nacelles identical with those of the Vickers Viscount 800 series.

Demonstrating its conviction that the AW650 would sell, the maker's initiated production with a speculative batch of ten aircraft, fabricated at the Armstrong Whitworth factories at Whitley and Baginton and assembled for flight test at Bitteswell. From Bitteswell, the first Argosy—as the AW650 was named after an interim period as the Freightercoach—was flown for the first time on January 8, 1959. Other aircraft followed quickly and the sixth of the initial batch of ten was in the air by the end of September.

Early interest in the Argosy was shown by Riddle Airlines with a provisional order for four placed on February 23, 1959. This was amended to a letter of intent for five in March, 1960, and the order was confirmed and increased to seven after Riddle entered a success-

ful bid to use the type on Logair operations in the U.S. These operations are scheduled daily freight services between USAF and Army bases to provide essential logistic support for military operations.

During 1960, six Argosies were involved in certification trials, leading to the full British and U.S. certificates of airworthiness being awarded on December 2. The first aircraft for Riddle was delivered to Miami on December 8, and went into service on the Logair routes on January 15, 1961. All seven aircraft had been delivered to Riddle by September and were each achieving about ten hours utilization a day.

On April 27, 1961, British European Airways signed a contract for the other three Argosies of the original batch of ten aircraft. After modification to BEA requirements, with windows blanked off, Rolamat loading system installed and different cockpit arrangements, flight trials of the first of these were resumed on September 24, 1961.

Delivery of the first BEA Argosy was made on November 6, and scheduled freight services began on December 1. All three aircraft were in service by January, 1962.

The ten initial aircraft are identified as AW650 Series 100s, with the Riddle aircraft designated Argosy 101 and those for BEA, Argosy 102. Further details are to be found in the Production Record.

In May, 1961, the makers announced that a further batch of Argosies was being laid down and that these would be of the Series 200 type with a redesigned box-spar wing of fail-safe structure. This version, for 1963 delivery, will have a gross weight of 90,000 lb., capacity payload of 31,000 lb., and a small increase in fuel capacity.

In addition to the civil AW650, the following variants of the design have been produced or projected.

AW651. A commercial version powered by two Rolls-Royce Tyne turboprops. Not being actively developed.

AW660. A military troop and supply transport. The RAF has ordered a total of fifty-six, the first of which (XN814) flew on March 4, 1961. The AW660 is distinguished by a new rear fuselage with "beaver tail" doors which can

---

1. *The first Argosy, G-AOZZ, later delivered to BEA.*
2. *The second Argosy, G-APRL, with temporary military-type rear doors.*
3. *One of the Riddle Argosy fleet, N6507R, in Logair markings.*
4. *BEA livery on the Argosy G-APRN.*

ARMSTRONG WHITWORTH **ARGOSY** (Great Britain)

be opened in flight to air-drop vehicles and supplies. Paratroop doors are on each side of the rear fuselage. The nose door is fixed and a nose radome is fitted. Provision is made for flight refuelling, and with four 2,680-e.h.p. Dart 101 (R.Da.10) engines the AW660 has a permissible overload weight of 105,000 lb.

The "beaver tail" doors were fitted experimentally to the civil Argosy 100 G-APRL, which first flew in this form on July 28, 1960. It was later converted back to normal standard.

AW670. A projected vehicle ferry version, with a new unpressurized fuselage wide enough for two rows of cars side by side, and having an upper deck for passengers. The undercarriage would be fixed and faired to simplify maintenance.

AW671. An "air bus" version of the AW670, with a similar fuselage providing a maximum of 114 high-density seats. Retractable undercarriage.

ARMSTRONG WHITWORTH 650 ARGOSY 100

Dimensions
Span: 115 ft. 0 in.
Overall length: 86 ft. 9 in.
Overall height: 27 ft. 0 in.
Gross wing area: 1,458 sq. ft.
Sweepback: Nil.
Internal cabin dimensions:
  Length (ex. flight deck): 46 ft. 8 in.
  Max. width: 120 in.; max. height: 80 in.
Max. usable floor area (ex. flight deck): 420 sq. ft.
Max. usable volume (ex. flight deck): 3,680 cu. ft.

Accommodation
Max. coach: 84 (5 abreast) at 32-in. pitch.

Powerplants
Four Rolls-Royce Dart 526 turboprops.
Take-off power each (I.S.A., s.l.): 2,100 e.h.p. at 15,000 r.p.m. Water injection system fitted.
Rotol four-blade reversing 11-ft. 6-in. propellers.

Weights and loadings
Basic operational: 48,500 lb.
Total fuel: 26,400 lb.
Mfrs. max. payload: 28,000 lb.
Max. take-off: 88,000 lb.
Max. landing: 84,000 lb.
Max. zero fuel: 77,500 lb.
Wing loading (max. t-o. wt.): 60.0 lb./sq. ft.
Wing loading (max. landing wt.): 55.5 lb./sq. ft.
Power loading (max. t.-o. wt.): 10.5 lb. e.h.p.

Performance
Optimum cost cruising speed (I.S.A.): 266 knots at 20,500 ft. and 79,000 lb.; consumption: 350 imp. gal./hr.
High-speed cruise, 288 knots at 25,000 ft.
Long-range cruise, 244 knots at 15,000 ft.
Approach speed (1.3Vs_0 at max. landing wt.): 118 knots.
Take-off field lengths (at max. t.-o. wt.):
  At I.S.A. at sea level: 4,900 ft.
  At I.S.A. +15° C. at sea level: 5,200 ft.
  At I.S.A. at 5,000 ft.: 6,350 ft.

Landing field length (max. landing wt.): 5,200 ft.
Range (take-off to landing), still air, no reserves, I.S.A., max. fuel (3,300 imp. gal.): 2,050 n. mi. with 13,100-lb. payload at 228 knots (mean) at 20,000 ft. (mean).
Range (take-off to landing), still air, no reserves, I.S.A., with max. payload, 750 n. mi. at 200 knots at 10,000 ft.

Structure. Semi-monocoque, open-ended fuselage. Two-spar mass boom wing in five sections. Twin tail booms.

Fuel system. Ten bag tanks between spars. Capacity 3,300 imp. gal. Two-point pressure refuelling at 300 g.p.m.

Undercarriage. Backwards retracting main oleos with twin wheels, and steerable twin nose-wheels. Disc brakes. Track, 32 ft. 6 in. Base, 25 ft. 5 in. Tyre pressure, 88 p.s.i.

Flying controls. All manual. Three-section ailerons with spring and geared tabs. Four-part elevator with geared spring and trim tabs. Rudders with spring and geared tabs. Smiths S.E.P.2 autopilot.

Flaps, air brakes. Double-slotted flaps in two segments each side. No air brakes.

Cabin conditioning. Three engine-driven blowers. Cabin differential, 5.5 p.s.i.

Hydraulics. Duplicated 2,500–3,000-p.s.i. system. Powered by three pumps to operate u/c, nose-wheel steering, brakes and flaps. Electric power pack for emergency use.

Electrics. Four 6-kW. engine-driven generators for 28-V. d.c.; two invertors for a.c. supply and three alternators for de-icing current.

De-icing. Two exhaust heat exchangers provide hot air for wings; electric heaters for tail unit, engine intakes and propeller. Nesa glass windscreen.

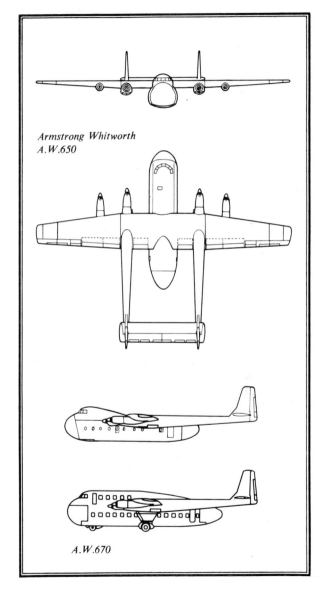

Armstrong Whitworth A.W.650

A.W.670

AVRO 748 (Great Britain)

This picture of the Avro 748 prototype G-APZV shows the original wing-tips; production aircraft have a small extension.

DEVELOPMENT OF THE Avro 748 feeder-liner began in 1957 as a result of a company decision to re-enter the civil aircraft market. Apart from civil versions of the Anson, there had been no commercial transport aeroplane bearing the Avro name since the Lancastrian, York and Tudor, developed during and immediately after the Second World War. By 1958, the project design team was concentrating upon a twin-engined aeroplane to carry about twenty passengers; with a gross weight of 18,000 lb. this required two 1,000-s.h.p. engines. High- and low-wing versions were studied, both under the type designation 748.

Airline reaction to these proposals was not wholly favourable, and resulted in a general scaling up of the design to take advantage of the thoroughly proven Rolls-Royce Dart powerplant. The Avro 748 then grew to DC-3 size and it became possible to offer DC-3 operators all the advantages of modern turbo-prop operations in an aircraft which could match their existing fleets unit for unit.

In January, 1959, the Hawker Siddeley Group decided to initiate production of the Avro 748 in this form, as a privately financed venture, and work began in the Manchester factory the following month on four proto-types—two for static testing and two for flight trials. Before the end of the year, the Indian Government purchased a licence to build the same basic type in India for military duties, and a second production line was laid down at the Indian Air Force Maintenance Base at Kanpur.

In structural design, the 748 was kept as simple and light as possible, to make for ease of inspection and repair away from base. Key factors in the design were an uninterrupted cabin space to provide maximum flexibility of seating arrangements, good airfield perform-ance and low operating cost.

The first prototype was assembled at Avro's Woodford, Cheshire, airfield, from which it made its first flight on June 24, 1960. Flight trials occupied the remainder of the year and most of 1961, with the second prototype joining the programme on April 10, 1961. These were both completed as Avro 748 Series 1s, this being the basic version with Dart 514 (R.Da.6) engines.

Early production aircraft are of a similar standard, and the first of these—one of three ordered by Skyways Coach-Air—made its first flight on August 30, 1961. Tropical trials had meanwhile been completed by the second prototype and a full British certificate of air-worthiness was awarded, on December 7, after Skyways crews had flown a 160-hr. route-proving programme on the first production 748. Features of the production aircraft are a small extension in wing span and an increase in engine power.

The second production aircraft, first flown on December 10, 1961, was the first of nine which had been ordered by Aerolineas Argen-tinas in March, 1960. This aircraft was formally delivered to representatives of the airline on January 11, 1962, and was flown out to Argentina a week later.

Another Series 1 produced by Avro was shipped out to India unassembled and was put together at the Kanpur factory, where it made its first flight, in Indian Air Force markings, on November 1, 1961. The Indian production version has been named Subroto.

Other variants of the Avro 748 have been designated as follows:

Avro 748 Series 2. An airframe identical with the Series 1 but powered by 1,910 s.h.p. Dart 529s (R.Da.7). These more powerful engines permit an increase in gross weight to 42,000 lb. for the same airport performance as that of the Series 1, or a better airfield perform-ance at equivalent weights. The higher weight also means longer sectors can be flown with full payload, which is increased in this version to 12,650 lb.

To obtain certification of the Series 2, the second prototype was converted to have the new engines and first flew in this form on November 6, 1961. Three Series 2s have been ordered by Aden Airways (VR-AAU to VR-AAW) for 1962 delivery and the Brazilian Air Force has ordered six.

---

1. *The first Avro 748 prototype, G-APZV.*
2. *The first two Avro 748s with the second, G-ARAY, nearest the camera.*
3. *Skyways livery on the first production 748, G-ARMV.*
4. *The second production 748 in Aerolineas Argentinas markings, LV-PIZ.*

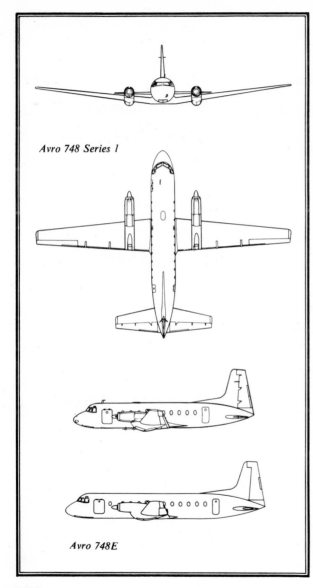

Avro 748 Series 1

Avro 748E

Avro 748M. This designation is used for the Indian production version, which will be based on the Series 2. The Kanpur factory will first assemble six 748s from Avro components, of which the first, as noted above, was a Series 1, flown on November 1, 1961.

Avro 748E. Derived from the Series 2, the 748E has a 6-ft. fuselage "stretch" with extra 3-ft. sections fore and aft of the wing. This makes it possible to carry forty-eight passengers in a comparable arrangement to the forty-seat Series 1 and 2, or sixty passengers in a high-density layout. The gross weight of this version remains 43,000 lb. but the payload increases to 13,570 lb.

Avro 748 Super E. For future development, this variant would combine the long fuselage of the 748E with 2,400-s.h.p. Dart R.Da.10s. These engines allow the gross weight to increase to 43,000 lb. and the payload to 14,000 lb. Performance is also improved, with the cruising speed going up to 270 knots.

AVRO 748 Series 1

Dimensions
Span: 98 ft. 6 in.
Overall length: 67 ft. 0 in.
Overall height: 24 ft. 10½ in.
Gross wing area: 810.75 sq. ft.
Sweepback: Nil.
Internal cabin dimensions:
    Length (ex. flight deck): 46 ft. 6 in.
    Max. width: 96.5 in.; max. height: 76 in.
Max. usable floor area (ex. flight deck): 296 sq. ft.
Max. usable volume (ex. flight deck): 1,990 cu. ft.

Accommodation
Normal tourist: 40 (4 abreast at 36-in. pitch).
Max. coach: 48 (4 abreast at 30-in. pitch).

Powerplants
Two Rolls-Royce Dart 514 (R.Da.6).
Take-off power each (I.S.A., s.l.): 1,740 e.h.p. at 14,500 r.p.m.
Rotol four-blade reversing 12-ft. propellers.

Weights and loadings
Basic operational: 22,614 lb.
Total fuel: 3,600 lb.
Mfrs. max. payload: 10,586 lb.
Max. take-off: 38,000 lb.
Max. landing: 36,300 lb.
Max. zero fuel: 33,600 lb.
Wing loading (max. t-o. wt.): 46.8 lb./sq. ft.
Wing loading (max. landing wt.): 44.7 lb./sq. ft.
Power loading (max. t-o. wt.): 10.6 lb./e.h.p.

Performance
Typical cruising speed (I.S.A.): 223 knots at 15,000 ft. and 34,000 lb.; consumption: 0.172 n. mi./lb.
Approach speed ($1.3 V_{s0}$ at max. landing wt.): 95 knots.
Balanced field lengths, max. t-o. wt., I.S.A. at sea level: 3,860 ft.
Landing field length (max. landing wt.): 3,600 ft.
Range (take-off to landing), still air, no reserves, I.S.A., max. fuel (1,140 imp. gal.): 1,520 n. mi. with 5,070 lb.
Range (take-off to landing), still air, no reserves I.S.A., with max. payload: 550 n. mi.

Structure. Circular-section fuselage. Two-spar wing in one piece each side, attaching to integral wing/fuselage centre section. Principal materials L65, L72, L73 and L77 aluminium alloys.

Fuel system. One integral tank each wing. Capacity 1,100 imp. gal. Gravity refuelling (pressure optional).

Undercarriage. Twin-wheel main legs retract forwards into nacelles. Steerable twin nosewheel folds forwards. Track, 24 ft. 9 in. Base, 20 ft. 8 in. Tyre pressure 4.0 p.s.i

Flying controls. All manual, with set-back hinges and horn-balanced rudder. Gear tab on each aileron and trim tab on one. Gear trim tabs on both elevators. Spring and trim tabs on rudder.

Flaps, air brakes. Fowler-type in one piece each side, with additional trailing edge tab to increase drag in landing. Electrically operated. No air brakes.

Cabin conditioning. Two engine-driven blowers. Differential, 4.2 p.s.i.

Hydraulics. 2,500-p.s.i. system powered by two pumps for u/c, brakes, nosewheel steering and optional propeller brakes (plus manual emergency pump for u/c).

Electrics. Primary 28-V. d.c. system for two engine-driven generators, plus two alternators for 115-V. a.c. and a third alternator for 200-V. a.c. for engine and propeller de-icing.

De-icing. Wing and tail by chordwise rubber boots energized by engine-bleed air.

BAC ONE-ELEVEN (Great Britain)

WORK ON THIS short-haul jetliner began in earnest in May, 1961, when an order for ten was placed by British United Airways. This event was preceded by several years of design work, undertaken in the first instance by Hunting Aircraft Ltd. In its earliest form, the project was known as the Hunting H-107 and was designed around two Orpheus engines.

Subsequently, Bristol Siddeley turbofans became available and by the end of 1959 the project was based on two 7,000-lb.s.t. B.E.61s. It then had a gross weight of 42,400 lb. and could carry forty-eight passengers over 700 naut. mile ranges.

After the Hunting company became part of the British Aircraft Corporation in 1960, the H-107 project continued as a BAC design, and was revised in the light of Vickers and Bristol commercial aircraft experience. As the BAC-107, it had a greater fuselage diameter for five-abreast seating, T-tailplane, small-diameter windows and 7,350-lb. BS75 turbofans.

At a gross weight of 56,500 lb., the BAC-107 will carry fifty passengers for 600–700 naut. miles.

The BAC-107 is still on offer to the airlines, for delivery when the BS75 engine becomes available. Meanwhile, the slightly larger One-Eleven (BAC-111) has been developed from it to take advantage of the Rolls-Royce RB163 Spey engines.

When the BUA order for One-Elevens was announced, BAC was also able to report a "letter of intent" from the U.S. local service airline Ozark for five. Subsequently, a similar option for six was taken by Frontier Airlines, and in October, 1961, Braniff Airways placed a firm order for six.

The first One-Eleven is to fly in the spring of 1963 with airline deliveries to follow in September, 1964. In addition to the version of the One-Eleven for which data are published below, there is a projected variant with a gross weight of 73,500 lb.

BAC ONE-ELEVEN

Dimensions
Span: 88 ft. 6 in.
Overall length: 93 ft. 8 in.
Overall height: 23 ft. 9 in.
Gross wing area: 980 sq. ft.
Sweepback: 20 deg. on quarter chord line.
Internal cabin dimensions:
    Length (ex. flight deck): 58 ft. 9 in.
    Max. width: 124 in.; max. height: 78 in.
Max. usable floor area (ex. flight deck): 539 sq. ft.
Max. usable volume (ex. flight deck): 3,150 cu. ft.

Accommodation
Typical mixed-class, 28F at 38-in. pitch and 29E at 36-in. pitch.
Max. coach: 74 (5 abreast at 34-in. pitch).
Volume of freight and baggage holds: 500 cu. ft.

Powerplants
Two Rolls-Royce RB-163-1 Spey.
Take-off power each (I.S.A., s.l.): 10,400 lb. at 12,490 r.p.m.
Sound suppressors and thrust reversers fitted. No water injection.

Weights and loadings
Basic operational: 42,800 lb.
Total fuel: 18,000 lb.

Max. payload: 14,000 lb.
Max. take-off: 68,250 lb.
Max. landing: 65,000 lb.
Max. zero fuel: 58,000 lb.
Wing loading (max. t-o. wt.): 69.6 lb./sq. ft.
Wing loading (max. landing wt.): 66.4 lb./sq. ft.
Power loading (max. t-o. wt.): 3.28 lb./lb. thrust.

Performance
Best cost. cruising speed (I.S.A.): 469 knots at 25,000 ft. and 66,000 lb.; consumption: 668 imp. gal./hr.; power (per engine): 3,540 lb. thrust.
Approach speed ($1.3V_{s_0}$ at max. landing wt.): 120 knots E.A.S.
Take-off field lengths required at max. t-o. wt.:
    At I.S.A. at sea level: 5,200 ft.
    At I.S.A. +15° C. at sea level: 6,350 ft.
    At I.S.A. at 5,000 ft.: 7,600 ft.
Landing field length required (at max. landing wt.): 5,810 ft.
Range (take-off to landing), still air, no reserves, I.S.A., max. fuel (2,250 imp. gal.): 1,835 n. mi. with 7,300 lb. payload at 435 knots (mean) at 30,000 ft. (mean).
Range (take-off to landing), still air, no reserves, I.S.A., with max. payload: 1,070 n. mi. at 435 knots (mean) at 30,000 ft. (mean).

BAC One-Eleven

BOEING 707 (U.S.A.)

In Western Air Lines' "Chieftain" livery, this Boeing 707-139 was originally built to Cubana order and later leased by the U.S. operator.

BOEING 707 (U.S.A.)

WHEN THE BOEING Company announced on August 30, 1952, that it was engaged in a $16 million privately financed programme to build a jet transport prototype, more than six years had already been spent in making design studies for this project. These studies, totalling over 150 separate designs, included some turboprop and turbojet transports based on the C-97 Stratofreighter and the B-47 Stratojet. Others were entirely original in concept and included some all-wing types.

The design eventually selected for construction as America's first jet transport was a close relative, aerodynamically, of the military B-47, with a wing sweep-back of 35 degrees and four podded engines. The early project studies were made under the designation of Model 473 for the jet variants and Model 367 for the turboprop types based on the Stratofreighter. As a measure to preserve secrecy, however, the design selected for development was called the Model 367-80 and as a result the prototype of the Boeing jet transport is now usually known as the "Dash Eighty". The production version of the design took the designation Model 707 in the normal sequence of maker's numbers.

From the original 367-80 prototype, a whole family of jet transports has been developed, and more of these had been sold to the world's airlines by the end of 1961 than of any other jet airliner. This family of variants —described in more detail below—goes under the Boeing 707 and Boeing 720 (q.v.) designations, the latter being an offshoot of the 707.

Sub-variants of the 707 have a series or "dash" number such as -120, -220, -320 or -420, indicating the size of the aircraft and the powerplant. Dash numbers with a "B" suffix indicate the use of turbofan engines. Each customer for the 707 is allocated a two-digit number, starting at 21, so that individual airline models carry dash numbers like -121, -124, -139 (for 707-120 series aeroplanes) or -321, -331 (for 707-320 series) or -436, -441 (for 707-420 series).

Dash Eighty. The Dash Eighty was intended from the start as a test aeroplane and demonstrator, and has given long and valuable service in this role. Smaller than the production-model Boeing 707s, it had a fuselage length of

119 ft. 6 in. and a fuselage width of 132 in. Both these dimensions were increased in the Boeing 707, but the wing remained substantially unchanged, as did the podded powerplant arrangement. Early model Pratt & Whitney JT3C-1 turbojets were fitted in the Dash Eighty, which made its first flight on July 15, 1954, in a distinctive brown and yellow finish and bearing the registration number N70700.

In the seven years from its first flight to July, 1961, the Dash Eighty completed twenty-four major test programmes involving 1,400 flying hours, and was continuing to undertake new tasks in 1962. After various aspects of the basic design had been satisfactorily demonstrated in flight, an early programme was conducted with the Dash Eighty as a refuelling tanker. A Boeing high-speed boom was added under the rear fuselage, together with a boom operator's station, and 350 flight hours were taken to test controls and demonstrate the boom and associated equipment. At this time also, the prototype acquired large cargo doors of the type later adopted on the KC-135A.

Another programme which temporarily changed the external appearance was the testing of the Bendix AMQ-15 weather reconnaissance system with a large nose radome and additional radomes and aerials on the fuselage. The Dash Eighty was also used to test various autopilot systems; radar-controlled landings; and interior and exterior sound levels with various types of suppressors on the engines.

For more than two years, the prototype flew with a JT3C-6 in one inner pod, a JT4A in the other inner pod, and the original

1. *The Boeing 367-80, America's first jet transport, in its original configuration.*
2. *For demonstrations to the USAF, dummy boom refuelling equipment was added to the Dash Eighty.*
3. *In another series of trials, Bendix AMQ-15 weather radar was fitted to the Dash Eighty.*
4. *Triple slotted landing flaps were tested on the Dash Eighty as part of a long development programme on high lift devices.*
5. *A fifth engine was added to the Dash Eighty as part of the development for the Boeing 727.*

JT3C-1s in the outer pods. Engine trials included a de-icing programme with a spray rig in front of the intakes of the starboard inner engine.

More than 250 Dash Eighty flight hours have been devoted to the development of high-lift devices on the wing, for introduction on later models of the 707 and 720. These developments included, first, plain and then slotted flaps on the leading edge; double and then triple slotted main flaps on the trailing edge; and a plain flap on the wing fillet. The aircraft was also flown for a time in 1960 with blown flaps, using air bled from the inboard engines.

A further important programme was started in 1961 in connection with the Boeing 727 (q.v.). For this, a fifth engine (a specially modified J57) was mounted on the side of the fuselage in the same manner as two of the Boeing 727's three engines. For the first trials, a "clam neck" tailpipe diverted the exhaust over the tailplane, but this was later replaced by clam-shell thrust reversing doors.

In April, 1962, the Dash Eighty resumed flight trials with four JT3D-1 turbofans replacing its JT3C engines, and a JT8D (Boeing 727 powerplant) was added on the rear fuselage later.

707-120. To meet specific airline requirements laid down after the Dash Eighty had been demonstrated, Boeing redesigned the fuselage to have adequate width (141 in.) for six-abreast seating (triple seats on each side of a centre aisle). With more powerful engines becoming available, the weight could also be allowed to increase beyond the Dash Eighty's 190,000 lb. and this in turn made possible a longer fuselage and a bigger fuel capacity.

As the 707-120, powered by 13,000-lb.-thrust (with water injection) JT3C-6 turbojets, this model was offered to the airlines in "long-body" (basic) and "short-body" versions, the respective fuselage lengths being 144 ft. 6 in.

---

6. *The second production 707-121.*

7. *Continental Airlines' first Boeing 707-124, N70773.*

8. *First Boeing 707-123 for American Airlines, N7501A.*

9. *A TWA Boeing 707-131, N731TW.*

and 134 ft. 6 in. The short-fuselage version eventually was ordered by Qantas, but all other customers for the 707-120 chose the longer body. The characteristics of the 707-120 were suited primarily to continental (e.g., U.S. domestic) operations.

The first airline order was announced on October 13, 1955, and was from Pan American World Airways for 20—only six of which were eventually taken by the airline as 707-120s. A production line was laid down at Boeing's Transport Division plant at Renton, near Seattle, Washington, and the first 707-120 was flown from the adjacent runway there on December 20, 1957. The second aircraft followed on February 3, 1958, and both were used, together with the third 707 off the line, for certification trials. Automatic wing leading-edge flaps of Krüger type were designed for the 707 during this time.

Provisional certification was obtained in August 15, 1958, and the first 707 was delivered to PAA on the same day. Late in August, Pan American began operating an experimental cargo service with 707s between New York and San Juan, Puerto Rico, and on October 26, 1958, the type was put into operation across the North Atlantic between New York and London, pending availability of the more suitable 707-320. This date marked the start of turbojet operation by U.S. airlines and with U.S. aircraft. Full FAA approval had been obtained on September 18, 1958.

The first U.S. domestic airline to order the 707-120 had been American Airlines and the first aircraft for this operator was delivered on October 23, 1958. Scheduled trans-continental operations by American began on January 25, 1959, but the 707 had already gone into domestic operations in the U.S. on December 10, 1958, between New York and Miami, for which National Airlines had chartered aircraft from Pan American.

The full list of airline customers for the 707-120 series, with dates of first service, is as follows: Pan American, six 707-121, October, 1958; National, 707-121 lease from Pan American, December, 1958; American, twenty-five 707-123, January, 1959; TWA, fifteen 707-131, March, 1959; Continental, five 707-124, June, 1959; Northeast, 707-131 lease from TWA, September, 1959; Avianca, 707-121

[continued on p. 17

BOEING 707 (U.S.A.)

BOEING 707-120/120B

Dimensions
Span: 130 ft. 10 in.
Overall length: 144 ft. 6 in.
Overall height: 42 ft. 0 in.
Gross wing area: 2,433 sq. ft.
Sweepback: 35 deg. on quarter chord line.
Internal cabin dimensions:
  Length (ex. flight deck): 104 ft. 10 in.
  Max. width: 140 in.; max. height: 91 in.
Max. usable floor area (ex. flight deck): 1,093 sq. ft.
Max. usable volume (ex. flight deck): 7,653 cu. ft.

Accommodation
Typical mixed-class: 40F (4 abreast at 40-in. pitch) and 66C (6 abreast at 36-in. pitch).
Max. high density: 179 (6 abreast).
Volume of freight and baggage holds: 1,660 cu. ft.

BOEING 707-120

Powerplants
Four Pratt & Whitney JT3C-6 turbojets.
Take-off power each (I.S.A., s.l.): 13,500 lb at 9,950 r.p.m., with water injection.
Sound suppressors and thrust reversers fitted.

Weights and loadings
Basic operational: 113,369 lb.
Max. (space limit) payload: 37,380 lb.
Max. fuel: 12,100 lb.
Max. take-off: 256,000 lb.
Max. landing: 185,000 lb.
Max. zero fuel: 170,000 lb.
Wing loading (max. t-o. wt.): 105.2 lb./sq. ft.
Wing loading (max. landing wt.): 76.0 lb./sq. ft.

Performance
Normal operating speed: Mach 0.884 above 24,900 ft.
High cruising speed (I.S.A.): 508 knots at 25,000 ft.
Long-range cruising speed (I.S.A.): 456 knots at 40,000 ft.
Approach speed ($1.3 V s_0$ at max. landing wt.): 139 knots.
Take-off field lengths (FAA requirements at max. t-o. wt.):
At I.S.A. at sea level: 10,200 ft.

---

10. *The Qantas short-fuselage Boeing 707-138, VH-EBA, with pre-delivery registration N31239.*

11. *Braniff Airways' first 707-227, N7071.*

12. *Second of the Intercontinentals, Pan American's 707-321, N715PA.*

13. *A Pan American 707-321, N723PA, on charter to Pakistan International.*

At I.S.A. +15° C. at sea level: 10,750 ft.
Landing field length (FAA requirements at max. landing wt.): 5,980 ft.
Range (take-off to landing), still air, no reserves, I.S.A., max. fuel (11,223 imp. gal.): 3,260 n. mi. with 43,850-lb. payload at 457 knots (mean) at 36,000 ft. (mean).
Range (take-off to landing), still air, no reserves, I.S.A., with max. payload: 2,895 n. mi.

BOEING 707-120B

Powerplants
Four Pratt & Whitney JT3D-1 turbofans.
Take-off power each (I.S.A., s.l.): 17,000 lb. at 10,200 r.p.m. (no water injection).
Thrust reversers fitted.

Weights and loadings
Basic operational: 120,734 lb.
Max. (space limited) payload: 37,380 lb.
Max. take-off: 258,000 lb.
Max. landing: 185,000 lb.
Max. zero fuel: 170,000 lb.
Wing loading (max. t-o. wt.): 106.1 lb./sq. ft.
Wing loading (max. landing wt.): 76.0 lb./sq. ft.

Performance
Normal operating speed: Mach 0.90 above 25,200 ft.
High cruising speed (I.S.A.): 540 knots at 25,000 ft.
Long-range cruising speed (I.S.A.): 470 knots at 40,000 ft.
Approach speed ($1.3 V s_0$ at max. landing wt.): 136 knots.
Take-off field lengths (FAA requirements at max. t-o. wt.):
At I.S.A. at sea level: 8,700 ft.
At I.S.A. +15° C. at sea level: 11,000 ft.
At I.S.A. at 5,000 ft.: 13,000 ft.
Landing field length (FAA requirements at max. landing wt.): 6,200 ft.
Range (take-off to landing), still air, no reserves, I.S.A., max. fuel (14,400 imp. gal.): 5,400 n. mi. with 18,830-lb. payload at 461 knots (mean) at 37,500 ft. (mean).
Range (take-off to landing), still air, no reserves, I.S.A., with max. payload: 3,725 n. mi.

Structure. Double-bubble monocoque fuselage; close-spaced frames and stringers 75ST and skins 24ST. Two-spar wing in two main panels joined by fuselage-width centre section; multi-panel tapered skins, 78ST and 24ST.

Fuel system. Three integral tanks in each wing and one in each root and up to five flexible cells in centre section. Capacity varies up to 14,493

lease from Pan American, October, 1960; Western, two 707-139 (first ordered by Cubana), June, 1960. The USAF also purchased three 707-153s for use by MATS as VIP transports with the designation VC-137A.

Several different operating weights and fuel tankages apply to these models, which are externally similar. The differences in fuel are accounted for by the number of flexible bay tanks in the centre section. Nominal tankages are: 707-121 and -153, 14,493 imp. gallons; 707-139, 13,808 imp. gallons; 707-131, 12,870 imp. gallons; 707-123, 12,920 imp. gallons and 707-124, 11,230 imp. gallons. Max. take-off weight of the 707-121, -123 and -131 is 246,000 lb.; of the 707-139, 215,000 lb. and of the 707-124 and -153, 256,000 lb.

The short-bodied 707 was ordered by Qantas in September, 1956, and deliveries of the fleet of seven began in June, 1959, certification having been obtained on June 24. This model carries the designation 707-138 in the

---

imp. gal. according to centre-section provision. Four-point pressure refuelling at 1,500 g.p.m.

Undercarriage. Four-wheel bogie main units, inwards retracting, and twin-wheel steerable nosewheel. Bendix wheels and brakes; Hydro-Aire anti-skid. Track, 22 ft. 1 in. Base, 59 ft. Tyre pressure 160 p.s.i.

Flying controls. Aerodynamically balanced manual low-speed ailerons, separate high-speed ailerons and differential spoilers. Manual elevators, boosted rudder. Electrically v.i. tailplane. Bendix PB-20D autopilot.

Flaps, air brakes. Double-slotted in two portions on each wing. Two segments of Krüger-type leading-edge flap. Two spoilers above and below each wing, and u/c as air brake up to 320 kt. I.A.S.

Cabin conditioning. AiResearch air-cycle system. Three (two on 720) turbocompressors powered by engine-bleed air. Two separate refrigeration systems. Cabin differential 8.6 p.s.i.

Hydraulics. Duplicated 3,000 p.s.i. system for u/c, brakes, flaps, spoilers, nosewheel steering and windscreen wipers.

Electrics. Four 30 kVA. three-phase alternators for 115-V. 400-cycle a.c. Rectifiers for 28-V. d.c., and 36 amp.-hr. battery.

De-icing. Wing and intakes by engine-bleed air; electric element mats on tail. Nesa glass windscreen.

-120 series and has a gross weight of 256,000 lb. and fuel capacity of 14,493 imp. gallons.

707-120B. To take advantage of the increased power and better fuel consumption of the turbofan version of the JT3 engine, Boeing developed the 707-120B. Chronologically this model followed the Boeing 720. Consequently, the -120B makes use of some of the aerodynamic refinements of the smaller aeroplane, in addition to the new engines.

The engines are 17,000-lb.s.t. (without water injection) JT3D-1s, similar to the JT3C-6s with the addition of a "fan" ahead of the compressor. Air from this fan is exhausted from an annulus round the forward part of the nacelle, which consequently is of distinctive appearance. No sound suppressors are fitted but both the main gas flow and that from the fan can be reversed for braking effect.

Additional features of the -120B are a "glove" on the wing leading edge, inboard of the inner pods, which decreases the effective wing thickness by increasing the chord, and thereby permits a 0.02 increase in the cruise Mach number; extra leading-edge flaps to improve the take-off performance, and a taller fin and a small ventral fin to improve handling in conditions of asymmetric thrust. The max. weight increased to 258,000 lb.

The first 707-120B was built to American Airlines order, as a replacement for a 707-123 lost on a training flight. This aircraft first flew on June 22, 1960, and the certificate of airworthiness was obtained on March 1, 1961. During 1960, American Airlines embarked on a programme to convert its twenty-four 707-123s to B standard and this work was completed early in 1962. The type entered service with American Airlines on March 12, 1961. TWA ordered eighteen 707-131Bs in 1961, for first delivery in March, 1962.

Qantas also ordered conversion of its fleet to turbofan standard, as 707-138Bs, and ordered four new aircraft of this type, to

14. *Boeing 707-331 N761TW in TWA colours.*
15. *An Air France Boeing 707-328, F-BHSB, after introduction of tail modifications.*
16. *Sabena's 707-329, OO-SJA.*
17. *South African Airways' 707-344, ZS-CKS, with modified tail.*

14

15

16

17

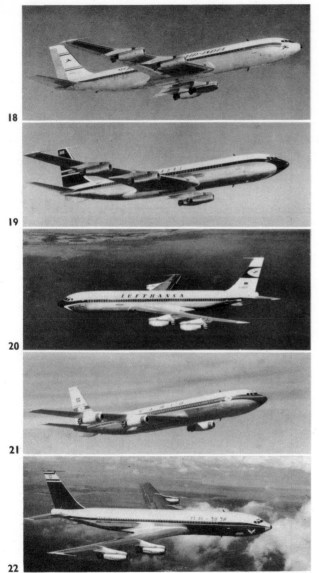

make a total fleet of eleven, of which one is on lease from Boeing. The first 707-138B was delivered in August, 1961, after certification on July 24, and went into service in September. It differs from the U.S. turbofan versions in having water injection. Gross weight is 258,000 lb., fuel capacity is 14,393 imp. gallons and permissible speed is Mach 0.91.

707-220. This version is similar to the -120 but has 15,800-lb.s.t. JT4A-3 or JT4A-5 engines to provide a better take-off performance in high temperatures and high altitudes. An order for five (Model 707-227) was announced by Braniff on December 1, 1955, and the first -220 flew on June 11, 1959. The type has been in service since December 20, 1959, having received its FAA Type certificate on November 5.

The gross weight of this version is 247,000 lb. and the fuel capacity is 12,870 imp. gallons.

707-320. To give the Boeing 707 a true intercontinental performance a larger version was designed; with the same fuselage cross section as the earlier model, it was lengthened by 8 ft. and the span was increased by 12 ft. Pratt & Whitney 15,800-lb.s.t. JT4A-3 turbojets were specified initially, and 16,800-lb. A-9 and A-10 versions of the same engine have been adopted by many operators of the -320. Known as the Intercontinental, the first 707-320 flew on January 11, 1959, and the certificate of airworthiness was obtained on July 15, 1959. Pan American began operating this model (707-321) on August 26, 1959, in the U.S. and on October 10 across the North Atlantic.

Apart from the increased size and fuel capacity, the 707-320 has additional wing leading-edge flaps, and since going into service most aircraft have been modified to have a taller fin and a ventral fin. These modifications

were developed on the 707-420 for BOAC (see below).

Airline orders and first service dates for the 707-320 are as follows: Pan American, twenty 707-321 and five 707-331, August, 1959; TWA, twelve 707-331, November, 1959; Air France, twenty-one 707-328, January, 1960; Sabena, seven 707-329, January, 1960; South African Airways, three 707-344, October, 1960; Pakistan International, one 707-321 lease from Pan American, March, 1960.

707-320B. Developed from the -320, the "B" model has 18,000-lb.s.t. JT3D-3 turbofan engines with nacelles as described above for the -120B. Other new features are low-drag wing-tips which increase the span to 145 ft. 8½ in., slotted leading-edge flaps and improved trailing-edge flaps. A 2,000-lb. increase is made in the maximum gross weight. The first 707-320B flew from Renton on February 1, 1962. Orders are as follows: Pan American, six 707-321B; TWA, five 707-331B; Air France, four 707-328B; Lufthansa, two 707-330B, and USAF, one 707-353B for Presidential use, designated VC-137C.

707-420. A version of the Intercontinental with Rolls-Royce Conway engines carries this designation. The first of five airlines to order this variant was BOAC, and the first of their aircraft flew on May 20, 1959. In the course of prolonged flight tests with four aircraft, improvements in the lateral control systems were developed; these comprised a 35-in. taller fin and fully power-boosted rudder, and a ventral fin which also prevents the nose from being lifted to too high an angle during take-off run.

A British certificate for the fully modified version was obtained on April 27, 1960, and BOAC started services in May. Other Conway-engined 707s were in service, with an American certificate granted on February 12, 1960, in March, 1960, but the modifications have been applied retrospectively to most 707s of all versions in service.

Orders for this version, and first service dates, are as follows: BOAC, eighteen 707-436, May, 1960; Air India, six 707-437, April, 1960; Lufthansa, five 707-430, March, 1960; Varig, two 707-441; El Al, three 707-458, July, 1961; Cunard Eagle, two 707-465, May, 1962.

---

18. *707-437, VT-DJT, in the original Air India colour scheme.*
19. *One of BOAC's 707-436s, G-APFE, with the tail modifications first developed for this variant.*
20. *Lufthansa 707-430, D-ABOF.*
21. *Varig, the Brazilian operator, has two 707-441s.*
22. *A Boeing 707-458 in El Al markings.*

[continued on p. 31

BOEING 707 (U.S.A.)

BOEING 707-320/320B (see also p. 31)

Dimensions
Span: 142 ft. 5 in. (-320B: 145 ft. 8½ in.).
Overall length: 152 ft. 11 in.
Overall height: 41 ft. 8 in. (-320B: 42 ft. 5½ in.).
Gross wing area: 2,892 sq. ft. (-320B: 2,942 sq. ft.).
Sweepback: 35 deg. on quarter chord line.
Internal cabin dimensions:
    Length (ex. flight deck): 111 ft. 6 in.
    Max. width: 140 in.; max. height: 91 in.
Max. usable floor area (ex. flight deck): 1,164 sq. ft.
Max. usable volume (ex. flight deck): 8,151 cu. ft.

Accommodation
Typical mixed class: 40F (4 abreast at 40-in. pitch) and 104C (6 abreast at 36-in. pitch).
Max. high density: 189 (6 abreast).
Volume of freight and baggage holds: 1,700 cu. ft.

BOEING 707-320

Powerplants
Four Pratt & Whitney JT4A-9.
Take-off power each (I.S.A., s.l.): 16,800 lb. at 9,135 r.p.m. (no water injection).
Sound suppressors and thrust reversers fitted.

Weights and loadings
Basic operational: 132,627 lb.
Max. fuel: 146,700 lb.
Max. (space limited) payload: 40,000 lb.
Max. take-off: 312,000 lb.
Max. landing: 207,000 lb.
Max. zero fuel: 190,000 lb.
Wind loading (max. t-o. wt.): 107.0 lb./sq. ft.
Wing loading (max. landing wt.): 71.7 lb./sq. ft.

Performance
Normal operating speed: Mach 0.887 above 24,900 ft.
High cruising speed (I.S.A.): 521 knots at 25,000 ft.
Long-range cruising speed (I.S.A.): 457 knots at 40,000 ft.
Approach speed ($1.3 Vs_0$ at max. landing wt.): 138 knots.
Take-off field lengths (FAA requirements at max. t-o. wt.):
    At I.S.A. at sea level: 10,200 ft.
    At I.S.A. +15° C. at sea level: 12,000 ft.
    At I.S.A. at 5,000 ft.: 14,400 ft.
Landing field length (FAA requirements at max. landing wt.): 7,200 ft.
Range (take-off to landing), still air, no reserves, I.S.A., max. fuel (19,380 imp. gal.): 5,900 n. mi. with 17,930 lb. payload at 453 knots (mean) at 38,400 ft. (mean).
Range (take-off to landing), still air, no reserves, I.S.A., with max. payload: 4,000 n. ni.

BOEING 707-320B

Powerplants
Four Pratt & Whitney JT3D-3 turbofans.
Take-off power each (I.S.A., s.l.): 18,000 lb. (with water injection).
Thrust reversers fitted.

Weights and loadings
Basic operational: 135,088 lb.
Max. (space limited) payload: 54,912 lb.
Max. take-off: 315,000 lb.
Max. landing: 207,000 lb.
Max. zero fuel: 190,000 lb.
Wing loading (max. t-o. wt.): 107.5 lb./sq. ft.
Wing loading (max. landing wt.): 70.5 lb./sq. ft.

Performance
High cruising speed (I.S.A.): 518 knots at 25,000 ft.
Long-range cruising speed (I.S.A.): 457 knots at 40,000 ft.
Approach speed ($1.3 Vs_0$ at max. landing wt.): 138 knots.
Take-off field length (FAA requirements at max. t-o. wt.):
    At I.S.A. at sea level: 9,800 ft.
Landing field length (FAA requirements at max. landing wt.): 6,420 ft.
Range (take-off to landing), still air, no reserves, I.S.A., max. fuel (19,830 imp. gal.): 7,000 n. mi. with 18,600 lb.
Range (take-off to landing), still air, no reserves, I.S.A., with max. payload: 4,980 n. mi.

Engineering summary—as on page 16 except:
Fuel system. Three integral tanks in each wing and one in each root and up to five flexible cells in centre section. Capacity varies up to 19,830 imp. gal., depending on centre-section provision.
Flaps, air brakes. Double-slotted in two portions on each wing, plus split fillet flaps. Four or six segments of Krüger-type leading-edge flap. 707-320B has slotted leading-edge flaps on outer sections and plain fillet flaps.

(see also p. 31)

23. *Cunard Eagle Airways' Boeing 707-465 G-ARWD before re-registration in Bermuda.*
24. *A Qantas 707-138B, VH-EBH, with turbofan engines.*
25. *The first Boeing 707 with turbofan engines, American Airlines' 707-123B, N7526A.*
26. *N760PA, the first 707-320B for Pan American.*

BOEING 720 (U.S.A.)

Photographed on a test flight over the Cascade mountains of Washington State, this Boeing 720-048 is in Avianca markings.

BOEING POLICY TOWARDS the development of its commercial jet airliners has been to offer a number of different versions of the same basic design, in order to meet as closely as possible the specific requirements of many different airlines. The result of this policy has been to produce a bewildering complex of type designations to cover three different sizes of aeroplanes, offered with several combinations of power plant and wing configuration.

The third "basic" member of the Boeing jet family was designed to operate most economically over shorter ranges than the 707-120 and 707-320. As the "baby" of the family, it was developed originally under the designation 707-020, but when it was first publicly announced in July, 1957, it was re-numbered Type 717. At this stage, the Boeing 717 was dimensionally identical with the short-body version of the 707-120, the overall fuselage length being 136 ft. 2 in.

With reduced fuel capacity and lower gross weight the 717 was capable of operating into and out of smaller airports than the 707. Further refinement of the basic design before the type went into production allowed a significant increase to be made in the permissible cruising speed.

In November, 1957, the designation was changed to 720. Customer numbers were in the 020 series (see explanation on page 13). Still dimensionally similar to the 707-120, the 720 introduced a "glove" which extended the wing leading edge forward, between the fuselage and the inner engine pylons. The effect of this was to increase the wing chord and, as the thickness of the wing section was unaltered, reduce ,the thickness/chord ratio. This aerodynamic refinement was primarily responsible for the Mach 0.02 increase in cruising speed at maximum cruise thrust.

The reduced fuel capacity led to various economies in structural weight, including a lighter undercarriage, all of which helped to keep down the gross weight. Lightweight versions of the Pratt & Whitney JT3C engines were used, without water injection. The lower weight in itself improved airfield performance, but this was further enhanced by the addition of leading-edge flaps on the outer wing sections, similar to those already used inboard of the engine pylons on the Boeing 707.

Other changes on the Boeing 720 included the deletion of the fin probe high-frequency radio antenna which is a distinctive feature of the 707, and deletion of one of the three turbo-compressors for the cabin conditioning system, together with the fresh-air intakes on the two outer engine pylons. The height of the fin was increased by 40 in., full power boosting was used in the rudder circuit and a small ventral fin was fitted—these items having been first developed for the Boeing 707-420 and subsequently introduced on earlier models of the 707.

United Air Lines became the first customer for the Boeing 720, with an order for eleven announced in November, 1957. Other orders followed steadily and by the end of 1961 totalled 120.

The first Boeing 720 was flown from Renton on November 23, 1959, and the FAA certification programme, in which three aircraft took part, began on January 18. FAA certification was obtained on June 30, 1960, and the type went into service with United Air Lines on July 5 and with American Airlines on July 31. According to customer requirement, the 720 is available at different gross weights (related to fuel capacity used) up to a maximum of 225,000 lb. At this weight, the fuel capacity is 11,303 imp. gallons, using the full integral wing tankage and two centre-section tanks. The minimum gross weight is 203,000 lb. with 9,875 imp. gallons of fuel. Basic powerplant for the 720 is the 12,000-lb.s.t. Pratt & Whitney JT3C-7, and the similar JT3C-12 is used by Eastern Air Lines.

The full list of Boeing 720 customers, with first service dates, is as follows: United, twenty-nine Model 720-022, July, 1960; American, ten Model 720-023, July, 1960; Irish International, three Model 720-048, December, 1960; Western, four Model 720-047, June, 1961; Braniff, four Model 720-027,

---

1. *The second Boeing 720-023 for American Airlines, N7528A.*
2. *United Airlines' 720-022, N7203U.*
3. *EI-ALA, one of three Boeing 720-047s used by Irish International Airlines.*
4. *An Eastern Air Lines' 720-053, N8702E.*

1961; Eastern, fifteen Model 720-025, September, 1961; Pacific Northern, two Model 720-062, May, 1962. In addition, the FAA purchased one Boeing 720-061, in Braniff configuration.

Boeing 720B. Like the Boeing 707-120B and -320B, the Boeing 720B has turbofan engines, and the improved performance which can be derived from their use. The 15 per cent. improvement in specific fuel consumption of the 17,000-lb.s.t. JT3D-1 engine gives the 720B a range with maximum fuel only a little below that of the 707-320 Intercontinental. Coupled with the aerodynamic improvements of the 720 airframe, these engines have made the 720B one of the most versatile of all the Boeing jet transports. The higher thrust allows an increase in gross weight to 234,000 lb. and, weight for weight, improves the airfield performance by a useful margin.

The first 720B flew on October 6, 1960, and was the second turbofan Boeing to fly. Like the first (the 707-120B), it was destined to join the fleet of American Airlines, which ordered fifteen Boeing 720Bs and, after they had been delivered, arranged for the conversion to the same standard of its ten 720s already in service. Certification of the 720B was completed on March 3, 1961, and American Airlines put the type into regular service, together with the 707-120B, on March 12.

The full list of customers for this variant, with first service dates, is as follows: American Airlines, fifteen Model 720-023B (plus ten converted 720-023s), March, 1961; Lufthansa, eight Model 720-030B, May, 1961; Avianca, two Model 720-059B, November, 1961; Ethiopian Air Lines, two Model 720-060B, 1962; Northwest, nine Model 720-051B (two on lease from Boeing), July, 1961; Pakistan International, three Model 720-040B, February, 1962; TWA, four Model 720-031B (lease from Boeing), August, 1961; El Al, two Model 720-058B, 1962; Continental, four Model 720-024B, 1962; Saudi Arabian Airlines, Model 720-068B, February, 1962.

BOEING 720/720B

Dimensions

Span: 130 ft. 10 in.
Overall length: 136 ft. 2 in.
Overall height: 41 ft. 6½ in.
Gross wing area: 2,433 sq. ft.
Sweepback: 35 deg. on quarter chord line.
Internal cabin dimensions:
    Length (ex. flight deck): 96 ft. 6 in.
    Max. width: 140 in.; max. height: 91 in.
Max. usable volume (ex. flight deck): 7,618 cu. ft.

Accommodation

Typical mixed class: 30F (4 abreast at 40-in. pitch and 60C (6 abreast at 36-in. pitch).
Max. high density: 140 (6 abreast).
Volume of freight and baggage holds: 1,390 cu. ft.

5. *Braniff's 720-027, N7077.*
6. *The first Boeing 720B, American Airlines' N7537A.*
7. *A Lufthansa 720-030B, D-ABOH.*
8. *A Western Airlines' 720-047B, N93143.*

BOEING 720

Powerplants

Four Pratt & Whitney JT3C-7 turbojets.
Take-off power each (I.S.A., s.l.): 12,000 lb. at 9,950 r.p.m. (no water injection).
Sound suppressors and thrust reversers fitted.

Weights and loadings

Basic operational: 105,040 lb.
Total fuel: 96,140 lb.
Max. (space limited) payload: 34,264 lb.
Max. take-off: 222,000 lb.
Max. landing: 175,000 lb.
Max. zero fuel: 142,000 lb.
Wing loading (max. t-o. wt.): 91.7 lb./sq. ft.
Wing loading (max. landing wt.): 71.9 lb./sq. ft.

Performance

Normal operating speed: Mach 0.906 above 23,400 ft.
High cruising speed (I.S.A.): 521 knots at 25,000 ft.
Long-range cruising speed (I.S.A.): 465 knots at 40,000 ft.

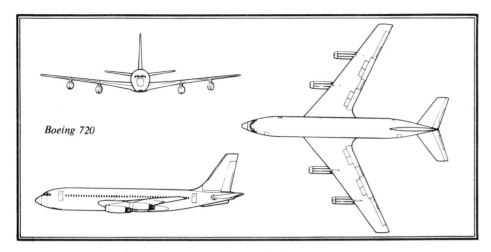

Boeing 720

Approach speed ($1.3 Vs_0$ at max. landing wt.): 126 knots.
Take-off field lengths (FAA requirements at max. t-o. wt.):
  At I.S.A. at sea level: 7,400 ft.
  At I.S.A. +15° C. at sea level: 8,300 ft.
  At I.S.A. at 5,000 ft.: 9,800 ft.
Landing field length (FAA requirements at max. landing wt.): 5,750 ft.
Range (take-off to landing), still air, no reserves, I.S.A., max. fuel (9,855 imp. gal.): 3,700 n. mi. with 18,270 lb. payload at 465 knots (mean) at 37,500 ft. (mean).
Range (take-off to landing), still air, no reserves, I.S.A., with max. payload: 2,600 n. mi.

BOEING 720B

Powerplants

Four Pratt & Whitney JT3D-1 turbofans.
Take-off power each (I.S.A., s.l.): 17,000 lb. at 10,200 r.p.m. (no water injection).
Thrust reversers fitted.

Weights and loadings

Basic operational: 109,580 lb.
Total fuel: 105,060 lb.
Max. (space limited) payload: 34,264 lb.
Max. take-off: 233,000 lb.
Max. landing: 175,000 lb.
Max. zero fuel: 156,000 lb.
Wing loading (max. t-o. wt.): 90.8 lb./sq. ft
Wing loading (max. landing wt.): 71.9 lb./sq. ft.

Performance

Normal operating speed: Mach 0.906 above 23,400 ft.
High cruising speed (I.S.A.): 540 knots at 25,000 ft.
Long-range cruising speed (I.S.A.): 470 knots at 40,000 ft.
Approach speed ($1.3 Vs_0$ at max. landing wt.): 134 knots.
Take-off field lengths (FAA requirements at max. t-o. wt.):
  At I.S.A. at sea level: 6,880 ft.
Landing field length (FAA requirements at max. landing wt.): 5,750 ft.
Range (take-off to landing), still air, no reserves, I.S.A., max. fuel (17,400 imp. gal.): 5,040 n. mi. with 19,300 lb. payload at 484 knots (mean) at 35,000 ft. (mean).
Range (take-off to landing), still air, no reserves, I.S.A., with max. payload: 3,885 n. mi.

Engineering summary—as on page 16 except:
Flaps, air brakes. Double-slotted in two portions on each wing, plus split fillet flaps. Six segments of Krüger-type leading-edge flap.

---

9. *N721US, Northwest Airlines' first 720-051B.*
10. *One of the four 720-031Bs leased by TWA from Boeing.*
11. *Pakistan International Airlines' first 720-040B, AP-AMG.*
12. *The Saudi Arabian Airlines' 720-068B, HZ-ACA.*

BOEING 727 (U.S.A.)

DESIGN STUDIES FOR a short-haul jet airliner were begun by Boeing during 1956, but it was not until December, 1960, that the design of the Boeing 727 had been finalized and manufacture began. In the intervening four and a half years, over sixty different possible configurations had been studied, covering two-, three- and four-engine layouts with the engines in various locations. Eventually, the required payload-range performance led to a choice of three engines and this, inevitably, resulted in a rear-engined layout closely similar to that which had already been adopted for the de Havilland Trident (q.v.).

For speed of production and lowest cost, Boeing chose a fuselage cross section which, from the cabin floor up, is identical with that of the Boeing 707 family. The engineering of the 727 in general follows established Boeing practices, but aerodynamically the design is more advanced, especially with respect of the high-lift devices on the wing. The leading- and trailing-edge wing flaps combine to give the 727 one of the highest maximum lift coefficients yet achieved.

Production of the 727 was initiated on December 5, 1960, after orders had been obtained from United Air Lines and Eastern Air Lines. The first 727 is scheduled to fly in the latter half of 1962, with certification expected in the final quarter of 1963. Airline service may begin before the end of the year, or early in 1964.

Orders for the Boeing 727 announced to date, and the constructor's numbers allocated, are as follows: United, forty Model 727-22 (18293–18332 inc.); Eastern, forty Model 727-25 (18252–18291 inc.); American, twenty-five Model 727-23 (18426–18450 inc.); Lufthansa, twelve Model 727-30 (18360–18371 inc.); TWA, ten Model 727-31.

An impression of the Boeing 727 in the markings of American Airlines.

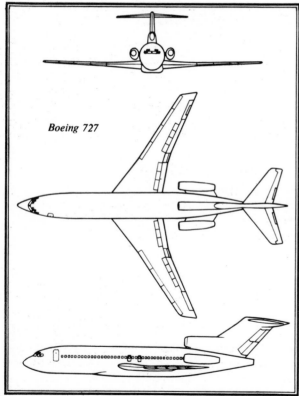

Boeing 727

BOEING 727

Dimensions

Span: 108 ft. 7 in.
Overall length: 135 ft. 5 in.
Overall height: 34 ft. 0 in.
Gross wing area: 1,650 sq. ft.
Sweepback: 32 deg. on quarter chord line.
Internal cabin dimensions:
    Length (ex. flight deck): 72 ft. 8 in.
    Max. width: 140 in.; max. height: 91 in.

Accommodation

Typical mixed class: 28F (4 abreast at 38-in. pitch) and 60C (6 abreast at 36-in. pitch).
Max. high density: 120 (6 abreast).
Volume of freight and baggage holds: 850 cu. ft.

Powerplants

Three Pratt & Whitney JT8D-1.
Take-off power each (I.S.A., s.l.): 14,000 lb. (no water injection).
Thrust reversers fitted.

Weights and loadings

Basic operational: 81,500 lb.
Total fuel: 46,900 lb.
Max. (space limited) payload: 24,000 lb.
Max. take-off: 142,000 lb.
Max. landing: 131,000 lb.
Max. zero fuel: 109,000 lb.
Wing loading (max. t-o. wt.): 86.0 lb./sq. ft.
Wing loading (max. landing wt.): 79.0 lb./sq. ft.

Performance

Best cost. cruising speed (I.S.A.): 515 knots at 25,000 ft. and 120,000 lb.
High cruising speed: 520 knots at 25,000 ft.
Approach speed ($1.3 V_{s_0}$ at max. landing wt.): 110 knots.
Take-off field lengths (FAA requirements at max. t-o. wt.):
    At I.S.A. at sea level: 5,250 ft.
    At I.S.A. +15° C. at sea level: 6,000 ft.
    At I.S.A. at 5,000 ft.: 7,600 ft.
Landing field length (FAA requirements, max. landing wt.): 4,980 ft.
Range (take-off to landing), still air, no reserves, I.S.A., max. fuel (5,830 imp. gal.): 3,000 n. mi. with 14,000 lb. payload at 445 knots (mean) at 35,000 ft. (mean).

BRISTOL **BRITANNIA** (Great Britain)

First turbine-powered airliner to operate regular schedules across the North Atlantic—the Britannia 312 of BOAC.

IN DECEMBER, 1946, the British Ministry of Supply drew up a draft specification for a new transport aeroplane to meet anticipated BOAC requirements. Known as the MRE (Medium Range Empire) requirement, or Specification 2/47, this envisaged an aeroplane with a design range of 2,500 miles carrying thirty-two passengers at 320 m.p.h. Despite the small accommodation wanted, this requirement eventually produced the 140-seat Britannia.

The Bristol Aeroplane Company was one of the five companies which submitted project designs to meet the 2/47 requirement, in April, 1947. This Bristol 175 was eventually accepted as the best of the new designs.

As first submitted, it was a 94,000-lb. aeroplane powered by four Bristol Centaurus piston engines, with a span of 120 ft. and a length of 114 ft. Even at this early date, however, Proteus turboprop engines were suggested as possible alternatives. During 1947, variations of the Bristol 175 design were studied at gross weights up to 107,000 lb., with greater fuel capacity and Centaurus, Proteus or Nomad compound diesel engines. The accommodation was increased to 40–44 seats in the capacious fuselage and the span went up to 130 ft.

From these alternatives BOAC chose, in October, 1947, an aeroplane which would weigh 103,000 lb. when fully laden, with a payload of 13,300 lb., a maximum of forty-eight seats and sufficient fuel to fly over the Karachi-Cairo stage, where prevailing headwinds and high temperatures produced critical conditions. Centaurus engines were planned initially although the Proteus was still being considered.

On July 5, 1948, the Ministry of Supply ordered three prototypes of the Bristol 175, on the understanding that BOAC would

purchase at least twenty-five. All three were to have Centaurus engines, but the last two would be suitable for ultimate conversion to Proteus power plants, with the third to be to full production standard. When BOAC ordered twenty-five examples of the Type 175 in November, 1948, they were to be built basically for Proteus engines, although the first six would have Centaurus to start with.

The span was increased again, to 140 ft., and accommodation for up to seventy-four passengers was now planned. In parallel with this work, Bristol began development of a long-range development suitable for trans-atlantic operations, for which the four-wheel bogie undercarriage was first adopted.

By January, 1951, progress with the Proteus engine led to cancellation of all plans for a Centaurus version and one of the prototypes was also cancelled. While the first prototype was being built, the weight was increased further to 140,000 lb. and additional tanks were put into the wing, these features being for introduction in the second prototype.

Britannia 101. This designation was applied to the two prototypes of the Series 100. The first (G-ALBO) flew at Filton on August 16, 1952, when it was powered by four Proteus 625 engines. The more powerful Proteus 705s were fitted in August, 1953. The second proto-type (G-ALRX) flew with these engines on December 23, 1953, and had the extra tankage and 140,000-lb. gross weight planned for production. After this prototype had been destroyed in a forced landing, G-ALBO was modified to continue flight tests at the higher weight. Subsequently, it was used for extensive engine development flying: with two Proteus 755s in the outer nacelles it first flew on April 17, 1956; with an Orion replacing one of the inner 705s it flew on August 31, 1956, and with a Proteus 765 replacing the other 705 it flew on November 5, 1957.

Britannia 102. Only fifteen of the original BOAC order for Britannias were eventually delivered in medium range configuration, with this designation. The first (G-ANBA) flew on September 5, 1954, and joined the prototype G-ALBO for certification flying. In the course of development flying prior to introduction on

---

1. *Britannia 101 prototype G-ALBO with the early nacelle arrangement.*
2. *The first production Britannia 102, G-ANBA, in BOAC colours.*
3. *BOAC Britannia 102, G-ANBO, on charter to Cathay Pacific.*
4. *The Britannia 301, G-ANCA, prototype of the long-fuselage version.*

BOAC routes, an engine icing problem was encountered which led to engine flame-outs when certain types of icing were experienced. Solving this problem required a massive development effort and delayed introduction of the Britannia into service.

BOAC eventually put the Britannia 102 into service on its routes to South Africa on February 1, 1957, and to Australia on March 1, 1957. Later, BOAC made aircraft of this type available to various of its associates, in whose colours they operated, including Central African, East African, Nigeria, Ghana, Malayan and Cathay Pacific Airways.

Britannia 252. Three examples of the long-fuselage, mixed-passenger-freight Britannia based on the Series 300 (see below) were ordered by the Ministry of Supply in February, 1955, and were to be made available to independent airlines for use on trooping. This scheme was not pursued and although the aircraft first flew with civil registrations they eventually were delivered to RAF Transport Command as Britannia C.Mk.2.

Britannia 253. In November, 1955, the RAF ordered six Britannias and later increased the total to twenty. These have the long fuselage of the Series 300, and have a large freight-loading door in the front fuselage plus a strengthened floor. The engines are Proteus 255 with water injection. All twenty were delivered as Britannias C.Mk.1s to RAF Transport Command at Lyneham for use by Nos. 99 and 511 Squadrons.

Britannia 300. Design work on a stretched version of the Britannia led to adoption of a 123-in. longer fuselage, Proteus 755 engines, 155,000-lb. weight and accommodation for 99 passengers. In mid-1953, BOAC took up an option on this variant, and later revised its contract to cover fifteen Series 100s and ten Series 300s. The latter quantity was reduced to eight in August, 1955, and finally to seven (see Britannia 301 and 302 below).

Britannia 301. This designation applied to the long-fuselage prototype, purchased by the Ministry of Supply. It first flew on July 31,

1956, and had the same wing and fuel capacity as the Series 100.

Britannia 302. Seven aircraft were laid down to this standard for BOAC but in 1956 the airline released them for sale abroad, and increased its order for Britannia 312s by seven aircraft. Two 302s were sold to Aeronaves de Mexico, which put them into service in December, 1957. The other five were ordered by Northeast Airlines of America in December, 1956, and were modified during production to have the fuel tankage developed for the Series 310. Thus modified they became Series 305, and when Northeast proved unable to finance its contract, they were sold to other operators as the Britannia 305, 307 and 309 (see below).

Britannia 305. Two of the ex-BOAC Northeast Britannias (see above) were ordered by the Argentine operator Transcontinental SA on August 24, 1959, and retained this designation. FAA type approval for the Britannia 305 was obtained on April 10, 1958.

Britannia 307. Two of the ex-BOAC Northeast Britannias ordered by Air Charter in July, 1958, took this type number. They were delivered in September, 1958, and subsequently joined the British United Airways fleet.

Britannia 309. The last of the original Series 300 aeroplanes modified to 305 standard was sold to Ghana Airways in 1960 (see also Britannia 319).

Britannia 310. To make the Britannia a genuine transatlantic aeroplane a version of the Series 300 was developed with integral tanks in the outer wing sections. Dimensionally the same as the Series 300 it had a gross weight of

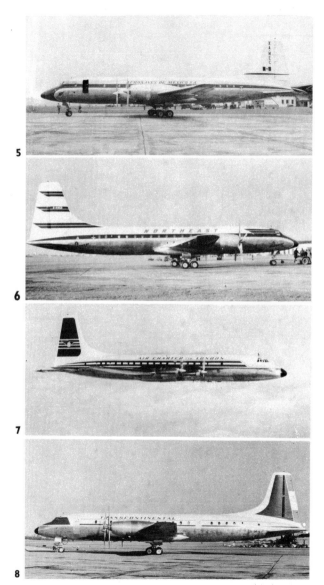

5. *A Britannia 302, XA-MEC, of Aeronaves de Mexico.*
6. *The Britannia 305, G-ANCD, temporarily painted in Northeast markings.*
7. *After delivery to Air Charter, Britannia G-ANCE was redesignated as a 307.*
8. *Transcontinental's Britannia 308, LV-GJB, first flew with the Class "B" registration G-14-1.*

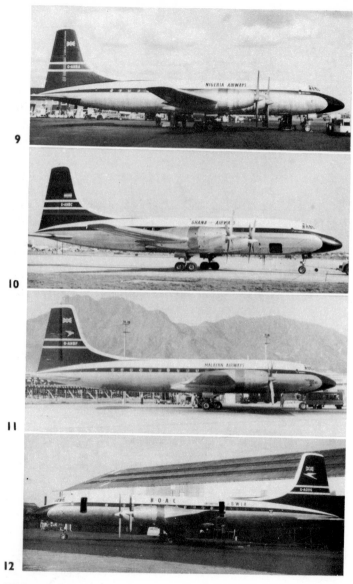

9

10

11

12

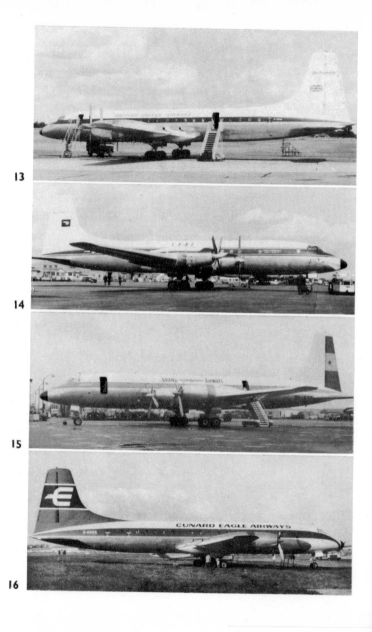

13

14

15

16

175,000 lb. (eventually increased to 185,000 lb.) and ten were ordered by BOAC in August, 1955, the contract later being increased to eighteen.

Britannia 311. After the prototype long-fuselage Britannia 301 (G-ANCA) had been lost in an accident, the first production Series 310 (G-AOVA) was diverted from the BOAC order and used for development flying. It first flew on December 31, 1956, and was sold to Ghana Airways as the Britannia 319 in 1960.

Britannia 312. This is the designation of the BOAC variant of the Series 310; eighteen were delivered to the airline, starting in September, 1957, and on December 19, 1957, BOAC in-augurated the first turboprop service across the North Atlantic (London–New York) with this type. Like the Britannia 102s, Britannia 312s have operated in the markings of several BOAC associates, including EAAC, CAA, Ghana and BWIA, and on charter to BUA.

Britannia 313. Four aircraft similar to the 312 purchased by El Al for transatlantic operation. The service was started three days after that by BOAC, on December 22, 1957. In 1962, El Al sold two of its Britannias to BUA.

Britannia 314. Canadian Pacific Air Lines ordered long-range Britannias on October 18, 1955, and six of the eight eventually acquired were of this type. The first was delivered on April 9, 1958, and operated its first Vancouver-Amsterdam service on June 1 (see Britannia 324).

Britannia 317. Two Series 310s were ordered by Hunting Clan in December, 1956, for charter flights and trooping. They were de-livered in December, 1958, and at first carried the markings of British and Commonwealth Shipping (Aviation), one of the parent Hunting Clan companies. After the Hunting Clan/Air-work merger, they were painted in BUA colours.

Britannia 318. Four Series 310s were purchased by Cubana, the Cuban airline, the first order being placed on May 24, 1957, and delivery taking place on December 15, 1958. During 1960 and 1961, one of these was leased to Cunard Eagle, and early in 1962 the same machine was leased to the Czech airline CSA.

Britannia 319. The first Series 310 G-AOVA (see also under Britannia 311) was sold to Ghana Airways in 1960, becoming the final Britannia sold, and was redesignated (see also Britannia 309).

Britannia 320. Towards the end of Britannia production, Bristol consolidated all the pro-gressive modifications and detail refinements, and redesignated the aeroplane as Series 320. Two aircraft for CPA carried designations in this series but four others—the 312 G-AOVT, the 313 4X-AGD and the 318s CU-P670 and CU-P671 were built to the same standard without redesignation.

Britannia 324. The final two Canadian Pacific aeroplanes, built to 320 standard, carried this designation. They were purchased from CPA by Cunard Eagle in 1961.

17

18

19

20

 9. *BOAC Britannia 102, G-ANBA, on charter to Nigeria Airways.*
10. *BOAC Britannia 102, G-ANBC, on charter to Ghana Airways.*
11. *BOAC Britannia 102, G-ANBF, on charter to Malayan Airways.*
12. *BOAC Britannia 312, G-AOVC, on charter to BWIA.*
13. *The Britannia 317, G-APNA, in BUA livery.*
14. *The BUA Britannia 317, G-APNB, in EAAC colours.*

15. *The Britannia 311 after delivery to Ghana Airways as 9G-AAH.*
16. *The Britannia 324, G-ARKB, purchased by Cunard Eagle from CPAL.*
17. *Ghana Airways' first Britannia, the Series 309, 9G-AAG.*
18. *The Britannia 311, G-AOVA, in prototype markings.*
19. *The El Al Britannia 313, 4X-AGB, later purchased by Cunard Eagle.*
20. *A Canadian Pacific Air Lines' Britannia 314, CF-CZC.*

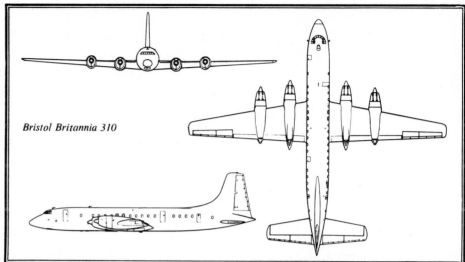

Bristol Britannia 310

BRISTOL BRITANNIA 320

Dimensions

Span: 142 ft. $3\frac{1}{2}$ in.
Overall length: 124 ft. 3 in.
Overall height: 37 ft. $4\frac{1}{2}$ in.
Gross wing area: 2,075 sq. ft.
Sweepback: Nil.
Internal cabin dimensions:
    Length (ex. flight deck): 86 ft. 3 in.
    Max. width: 139 in.; max. height: 80 in.
Max. usable floor area (ex. flight deck): 883 sq. ft.
Max. usable volume (ex. flight deck): 5,877 cu. ft.

Accommodation

Typical mixed class: 28F (4 abreast at 40-in. pitch) and 66C (6 abreast at 39-in. pitch).
Max. high density: 139 (6 abreast at 34-in. pitch).
Volume of freight and baggage holds: 819 cu. ft.

Powerplants

Four Bristol Siddeley Proteus 765.
Take-off power each (I.S.A., s.l.): 4,450 e.s.h.p. at 11,775 r.p.m. No water injection.
De Havilland four-blade 16-ft. reversing Hydromatic propellers.

Weights and loadings

Basic operational: 93,631 lb.
Total fuel: 68,376 lb.

Max. (space limited) payload: 30,000 lb.
Max. take-off: 185,000 lb.
Max. landing: 137,000 lb.
Max. zero fuel: 128,000 lb.
Wing loading (max. t-o. wt.): 89.2 lb./sq. ft.
Wing loading (max. landing wt.): 67.0 lb./sq. ft.
Power loading (max. t-o. wt.): 10.4 lb./e.s.h.p.

Performance

Best cost. cruising speed (I.S.A.): 349 knots at 21,000 ft. and 150,000 lb.; consumption: 765 imp. gal./hr.
Approach speed ($1.3 V s_0$ at max. landing wt.): 115 knots.
Take-off field lengths (to 1955 BCAR, at max. t-o. wt.):
    At I.S.A. at sea level: 8,000 ft.
    At I.S.A. +15° C. at sea level: 9,600 ft.
    At I.S.A. at 5,000 ft.: 11,700 ft.
Landing field length (to 1955 BCAR, at max. landing wt.): 6,000 ft.
Range (take-off to landing), still air, no reserves, I.S.A., max. fuel (8,547 imp. gal.): 4,500 n. mi. with 22,993 lb. payload at 310 knots (mean) at 28,000 ft. (mean).
Range (take-off to landing), still air, no reserves, I.S.A., with max. payload: 3,100 n. mi. at 349 knots (mean) at 21,000 ft. (mean).

21. *Later used by BUA, the Britannia 317, G-ANPA, originally carried British Commonwealth markings.*
22. *A Cubana Britannia 318, CU-P668.*
23. *Eagle Airways (later Cunard Eagle) leased the Britannia 318, CU-T668, from Cubana as G-APYY.*
24. *After its return to Cubana by Cunard Eagle, this Britannia 318 was leased by CSA for transatlantic operations as OK-MBA.*

BRISTOL BRITANNIA (Great Britain)

Structure. Semi-monocoque circular-section fuselage. Box-spar wing in two main units joined at fuselage centre line. Detachable wing tips.

Fuel system. One integral and two bag tanks in each wing and one bag tank in centre section. Capacity 8,580 imp. gal. Two-point pressure refuelling at 300 g.p.m.

Undercarriage. Four-wheel bogie main and twin-wheel steerable nose units. Dunlop wheels, dics brakes and Maxarets. Track, 31 ft. Base, 42 ft. 1 in. Tyre pressure 140 p.s.i.

Flying controls. Aerodynamic full-servo type operated by tabs on each surface. Spring feel for ailerons and rudder; *q* feel for elevators. Smiths S.E.P.2 autopilot.

Flaps, air brakes. Double-slotted Fowler type, electrically operated. No air brakes.

Cabin conditioning. System fed by engine-bleed air through heat exchangers. Differential, 8.3 p.s.i.

Hydraulics. Primary 4,000-p.s.i. system for u/c, brakes, nosewheel steering and control-system ground locks. Separate hydraulic system for *q* feel and control damping.

Electrics. Four 47-kVA. three-phase generators for 208-V. a.c. and transformer rectifiers for 112-V. and 28-V. d.c. Five invertors for radio and instruments fed from 112-V. busbar.

De-icing. Wing and engine intakes by hot gas from turbine and by electrically heated spray mats on tailplane, fin and elevator horn balances. Propellers and windscreen by electric elements.

BOEING 707 (*continued from page 19*)

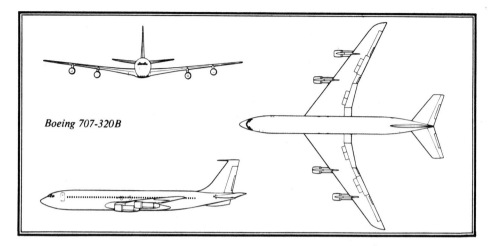

Boeing 707-320B

Boeing 707-320C. This designation refers to a convertible cargo-passenger version of the Boeing 707-320B. Its special features are a large freight-loading door, measuring 91 in. by 134 in., in the forward port fuselage, and an increased gross weight of 327,000 lb. A movable bulkhead in the fuselage makes it possible for the 707-320C to carry various combinations of passenger and freight loads.

In April, 1962, Pan American announced an order for two 707-321Cs for delivery early in 1963. Apart from the special features for freight carrying, they will be identical with the five 707-321Bs ordered by Pan American.

Prior to development of the -320C, Boeing had examined a number of other freight projects based on the Boeing 707 design. Many of these were in the Boeing 735 design series and included pure freighters with swing-tail loading and turbojet or turbofan engines.

707-520B. A projected development of the Boeing 707-320B, this variant is being prepared for marketing in 1963. With all the advanced features of the -320B it combines a 12-ft. longer fuselage for greater passenger loads, greater fuel capacity and 21,000-lb.s.t. JT3D-5A turbofans. Up to the spring of 1962, no orders had been placed for this version, which is almost certainly the ultimate "stretch" of the basic Boeing 707 design.

N746TW, the first of eighteen Boeing 707-131Bs for TWA.

CANADAIR **FORTY FOUR** (Canada)

Fairings on the rear fuselage cover hinges for the swing-tail feature of this Slick Airways Canadair Forty Four.

CANADAIR FORTY FOUR (Canada)

THE CANADAIR FORTY FOUR or CL-44 is the largest pure freighter so far put into service by the airlines, and the first to feature a "swing tail" for ease of loading. This arrangement, by which the entire rear fuselage section, including the tail unit, can be swung to one side, had previously been proposed by various manufacturers and built into one or two prototypes, but the Forty Four is the first aeroplane to go into regular service with it.

Work on the design was originated by Canadair in 1956 to meet an RCAF requirement for a long-range troop and freight transport. The Canadian company already had a licence from Bristol to build versions of the Britannia and was in production with the maritime reconnaissance Argus based on the British design. The Britannia was also taken as the basis for the new transport design, therefore, although eventually comparatively little was left beyond the wing torsion box. Several projects were studied under the CL-44 designation, with a longer fuselage than the Britannia and various engines, including R-3350 piston radials, and Pratt & Whitney T-34, Bristol Orion and Rolls-Royce Tyne turboprops. In 1958, the RCAF placed an order for eight CL-44s with the Orion engines.

A few months later, development of the Orion—which had been projected for advanced Britannia variants—was cancelled, and the Canadair transport had to be redesigned to take the Rolls-Royce Tyne turboprops. In this form it became the CL-44D, with the RCAF version known as the CL-44-6 to the makers and as the CC-108 Yukon to the RCAF. The military order was eventually increased to twelve and the first of these (15501) was flown for the first time on November 15, 1959.

Deliveries to the RCAF were completed during 1961. Subsequent to the first flight, the first aircraft was renumbered 15921 and the remainder took consecutive numbers to 15932. The first two were later modified to have a more comfortable interior for special duties and renumbered again, as 15555 and 16666 respectively.

The RCAF CL-44s have large freight loading doors fore and aft in the fuselage side. The swing tail feature was developed for the special requirements of civil operators and, as the CL-44D-4, this version was ordered by Flying Tiger, Seaboard World and Slick. Canadair laid down a company-financed prototype which was built as the ninth aircraft on the military production line and first flew on November 16, 1960.

Offered originally with a gross weight of 205,000 lb., the CL-44D-4 has been developed to have a 5,000-lb. higher take-off weight and a number of other performance improvements. Deliveries began to Flying Tiger Line on May 31, 1961, and this company has put the type into service on its military charter flights between the U.S. west coast and Japan. Seaboard World Airlines received its first CL-44D-4 on June 20, 1961, and subsequently introduced the type on scheduled transatlantic freight services. The third operator of the type is Slick, which received its first on January 17, 1962. All seventeen aircraft purchased by these three operators to date are of basically similar type, with small differences for domestic or international operations. Another ten were laid down on speculation in 1961, jointly financed by the Canadian Government and Canadair.

Several further developments of the Forty Four were projected by Canadair during the production of the initial version. These included the following:

CL-44E. A variant of the "D" to take full advantage of the potential weight growth of the airframe and possible power growth of the Tyne R.Ty.12. Gross weights up to 234,000 lb. were projected to provide greater payload or greater range.

CL-44F. This designation covered a commercial version of an airborne early-warning development of the Forty Four offered to the USAF.

---

1. *The prototype swing-tail Canadair CL-44, CF-MKP-X.*
2. *Slick Airways' first CL-44, N6035A.*
3. *A Flying Tiger Line's CL-44.*
4. *A Seaboard World Airlines' CL-44 with pre-delivery registration CF-MKP-X.*

5

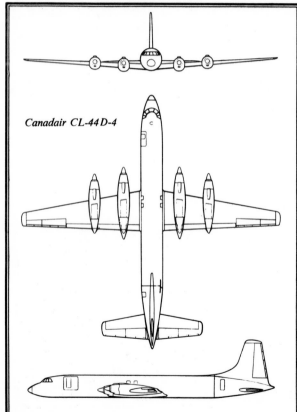

Canadair CL-44D-4

CL-44G. Similar to the "D" with the floor level aft of the main spar dropped by two feet. This was to facilitate loading of vehicles and other bulky material and was a commercial version of a swing-tail freighter offered to the RCAF in 1959.

---

5. *A Canadair CC-108 Yukon in RCAF markings.*

CANADAIR CL-44D

Dimensions

Span: 142 ft. 3½ in.
Overall length: 136 ft. 8 in.
Overall height: 38 ft. 7 in.
Gross wing area: 2,075 sq. ft.
Sweepback: Nil.
Internal cabin dimensions:
   Length (ex. flight deck): 98 ft. 1 in.
   Max. width: 137 in.; max. height: 81.5 in.
Max. usable floor area (ex. flight deck): 1,080 sq. ft.
Max. usable volume (ex. flight deck): 6,239 cu. ft.
Volume of under-floor freight holds: 1,109 cu. ft.

Powerplants

Four Rolls-Royce Tyne R.Ty.12 Mk. 515 turbo-props.
Take-off power each (I.S.A., s.l.): 5,730 e.s.h.p. at 15,250 r.p.m.
De Havilland four-blade 16-ft. reversing Hydromatic propellers.

Weights and loadings

Basic operational: 90,000 lb. (domestic cargo).
Total fuel: 81,500 lb.
Max. (weight limited) payload: 65,000 lb.
Max. take-off: 210,000 lb.
Max. landing: 165,000 lb.
Max. zero fuel: 155,000 lb.
Wing loading (max. t-o. wt.): 101.2 lb./sq. ft.
Wing loading (max. landing wt.): 79.9 lb./sq. ft.
Power loading (max. t-o. wt.): 9.1 lb./e.s.h.p.

Performance

Best cost. cruising speed (I.S.A.): 349 knots at 20,000 ft. and 165,000 lb.); consumption: 0.41 lb./t.e.h.p./hr.
Approach speed (1.3V_{s0} at max. landing wt.): 132 knots.

Take-off field lengths (FAA requirements at max. t-o. wt.):
   At I.S.A. at sea level: 7,350 ft.
Landing field length (FAA requirements, max. landing wt.): 5,980 ft.
Range (take-off to landing), still air, no reserves, I.S.A., max. fuel (8,458 imp. gal.): 5,680 n. mi. with 37,440-lb. payload at 343 knots (mean) at 25,000 ft. (mean).
Range (take-off to landing), still air, no reserves, I.S.A., with max. payload: 3,550 n. mi.

Structure. Circular-section monocoque fuselage. Two-spar wing in two main units joined at fuselage centre-line. Stiffened, stress-carrying upper and lower skins. Principally 75ST and 24ST.

Fuel system. Six-bay integral tanks in wings; capacity 8,520 imp. gal. Optional six-cell bag tank in centre-section; capacity 1,620 imp. gal. Pressure refuelling at 500 g.p.m.

Undercarriage. Four-wheel bogie main units, backwards retracting, and steerable twin-wheel nosewheel. Hydraulic brakes and anti-skid units. Track, 31 ft. Base, 49 ft. 11 in. Tyre pressure 160 p.s.i.

Flying controls. Fully manual, with servo-operation by tabs on ailerons, elevators and rudder. Spring-loading in aileron and rudder circuits and hydraulic artificial feel system for elevator. Bendix PB-20H autopilot.

Flaps, air brakes. Double-slotted flaps in one portion on each wing. No air brakes.

Cabin conditioning. Two engine-driven compressors. Differential, 8.33 p.s.i.

Hydraulics. Two engine-driven pumps power main 3,000 p.s.i. system for u/c, brakes, nose-wheel steer, cargo door and swing-tail operation, with emergency compressed-air system for u/c and brakes.

Electrics. Three engine-driven 40-kVA. three-phase 400-cycle alternators for 115-V. a.c., plus variable frequency a.c. from three 60-kVA. generators and 28-V. d.c. through rectifiers.

De-icing. Wing by ram-air passed over engine tail-pipes; tail unit by hot air from dual heater system; electric de-icing for intakes, propellers, screen and elevator horn balances.

CANADAIR 540 (Canada)

One of three Convair airframes converted by Canadair, CF-LMA was used as a company demonstrator.

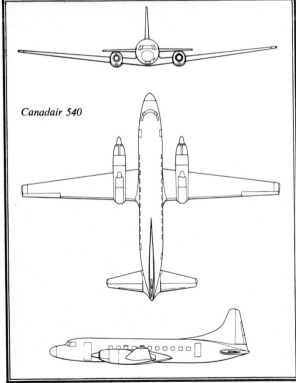

Canadair 540

SEVERAL DESIGNATIONS and names apply to this aircraft, which is a Convair 340 or 440 with its two piston engines replaced by Napier Eland turboprops. The first such conversion was made by Napier in England, using a Convair 340 (N8458H) purchased in the U.S.A. With Eland NEl.1s and re-registered G-ANVP it first flew on February 9, 1955. This aircraft eventually was leased to Allegheny Airlines and put into service on routes out of Philadelphia, starting in July, 1959, as N340EL.

A second similar conversion was made at Santa Monica for Napier by PacAero, and was widely demonstrated as N440EL before being sold as an executive transport to Butler Aviation. Five more conversions purchased by Allegheny were made by AiResearch and

ABOVE—*One of the five Convair 540s used by Allegheny Airlines until the spring of 1962.*

were registered N540Z to N544Z respectively. They were withdrawn from service early in 1962 when development of the Eland was discontinued.

In 1958, the RCAF ordered ten Eland Convairs, under the designation CC-109. These were built by Canadair as new aircraft at Toronto, with the company type number CL-66. The first (11151) flew on January 7, 1960, and all ten were delivered between July 15, 1960, and February 10, 1961. Civil versions were offered as Canadair 540s, but none was built as such. Canadair made three conversions of existing Convair airframes before delivering the CL-66s, and these were known variously as Canadair 540s and Convair 540s. The first two were lent to the RCAF for training and later went to Quebecair as CL-MKO and CF-LMA. These were returned to Canadair early in 1962, and the company also uses the third conversion, CF-LMN.

CANADAIR **540**

Dimensions
Span: 105 ft. 4 in.
Overall length: 79 ft. 2 in.
Overall height: 28 ft. 2 in.
Gross wing area: 920 sq. ft.
Sweepback: 4.3 deg. on quarter chord line.
Internal cabin dimensions:
 Length (ex. flight deck): 54 ft. 10 in.
 Max. width: 107 in.; max. height: 79 in.
Max. usable floor area (ex. flight deck): 396 sq. ft.
Max. usable volume (ex. flight deck): 2,515 cu. ft.

Accommodation
Normal first-class: 48 (4 abreast at 38-in. pitch).
Max. high density: 58 (4 abreast at 32-in. pitch).
Volume of freight and baggage holds: 402 cu. ft.

Powerplants
Two Napier Eland 504A turboprops.
Take-off power each (I.S.A., s.l.): 3,500 e.s.h.p. at 12,500 r.p.m.
De Havilland four-blade 13-ft. 6-in. propellers.

Weights and loadings
Basic operational: 34,061 lb.
Total fuel: 13,614 lb.
Max. (weight limited) payload: 12,939 lb.
Max. take-off: 53,200 lb.
Max. landing: 50,670 lb.

Max. zero fuel: 47,000 lb.
Wing loading (max. t-o. wt.): 57.9 lb./sq. ft.
Wing loading (max. landing wt.): 55.1 lb./sq. ft.
Power loading (max. t-o. wt.): 7.6 lb./e.s.h.p.

Performance
Best continuous cruising speed (I.S.A.): 278 knots at 14,000 ft. and 48,000 lb.; consumption: 309 imp. gal./hr.
Long-range cruise, 225 knots at 20,000 ft. at 48,000 lb. (mean).
Approach speed ($1.3 V_{s_0}$ at max. landing wt.): 130 knots.
Take-off field lengths (FAA requirements at max. t-o. wt.):
 At I.S.A. at sea level: 4,550 ft.
 At I.S.A.+15° C. at sea level: 5,250 ft.
 At I.S.A. at 5,000 ft.: 6,175 ft.
Landing field length (FAA requirements, max. landing wt.): 4,020 ft.
Range (take-off to landing), still air, no reserves, I.S.A., max. fuel (1,702 imp. gal.): 1,680 n. mi. with 5,675 lb. payload at 278 knots (mean) at 20,000 ft. (mean).
Range (take-off to landing), still air, no reserves, I.S.A., with max. payload: 774 n. mi. at 278 knots (mean) at 20,000 ft. (mean).

Structure. Circular-section semi-monocoque fuselage. Two-spar wing in one piece each side.

Fuel system. One integral tank in each wing outboard of nacelles. Capacity 1,700 imp. gal. Gravity refuelling.

Undercarriage. Forwards retracting main oleos. Dual main and steerable nosewheels. Track, 25 ft. Base, 26 ft. 1¾ in.

Flying controls. Fully manual system, with mechanical trim tabs on each aileron, elevator and rudder. Sperry A-12 autopilot.

Flaps, air brakes. Fowler-type flaps. No air brakes.

Cabin conditioning. AiResearch compressor on starboard engine. AiResearch air-cycle cooling. Differential, 4.16 p.s.i.

Hydraulics. 3,000 p.s.i. system for u/c, brakes, nosewheel steering, flaps, airstairs and ground blower. Two engine-driven pumps and emergency electric pump.

Electrics. Two engine-driven 28-V. d.c. generators and two 35-kVA. engine-driven three-phase 208-V. a.c. generators.

De-icing. Leading edges by engine-exhaust air. Nesa glass windscreen. Electric cycle propeller de-icing; Spraymat on spinners.

CONVAIR 880 U.S.A.

THE CONVAIR DIVISION of the General Dynamics Corp. became the third major U.S. manufacturer committed to the production of a large jet transport in September, 1956, when orders were obtained from TWA and Delta. The aircraft in question eventually went into service in May, 1960, as the fastest of the trio of American four-jet airliners, all of which closely resemble each other (see Boeing 707, pp. 12–19 and 31, and Douglas DC-8, pp. 55–62).

Convair development of a commercial jet transport began a little later than similar work by Boeing and Douglas, with a detailed market study. This study led the company to concentrate on an aeroplane of rather smaller capacity but higher performance than the 707 and DC-8, aimed particularly at U.S. domestic operators. The general configuration was a low-wing aeroplane with four podded engines and 35 degrees of wing sweep-back. After some initial indecision about the powerplant, the General Electric J79 was selected and the design became firm with these engines in January, 1956.

When the TWA and Delta orders were announced on September 10, 1956—for thirty and ten respectively—the project was known as the Convair Skylark, but this name was later changed to Golden Arrow. The new name matched a scheme to give the exterior skin a golden tint by anodizing, but both this scheme and the name were eventually dropped. The project was identified as the Model 22 in the manufacturer's design series and became the Convair 600 for publicity purposes, this number being derived from the planned cruising speed of 606 m.p.h. This speed expressed as feet per second produced the aircraft's ultimate designation, Convair 880.

In the course of design development, several changes were made, including lowering the tailplane from the base of the fin to the fuselage, and the gross weight grew from 172,500 lb. to 178,500 lb. Design refinements allowed the max. cruising speed to go up from 609 m.p.h. to 615 m.p.h.

Before certification the gross weight went up again, to 184,500 lb. The fuselage size was limited to allow a maximum of five seats across the width of the cabin, in contrast with the six-abreast seating in other types.

During 1957, additional orders for the Convair 880 were placed by two South American operators—Transcontinental S.A. and REAL Aerovias each buying four. Transcontinental subsequently was unable to finance this contract, however, and the REAL order was dropped in favour of Convair 990s (q.v.).

While a production line was being established at Convair's San Diego factory—in keeping with normal U.S. practice, no prototypes were built—design developments of the Convair 880 included two new versions. One of these was planned as an international model, with extra fuel in the centre section and leading edge flaps. This was identified as the Convair Model 31 and eventually became known as the 880-M. The second development (Model 30) was a more extensive redesign with a longer fuselage, turbofan engines and extra fuel, which became the Convair 990.

The first Convair 880 flew for the first time on January 27, 1959, at San Diego. This particular aircraft was built as number one of the TWA order but eventually was retained by Convair as a development aircraft and demonstrator. The second aircraft, also for TWA, was first used for static tests and did not fly until much later. Next to fly, therefore, was the third production Convair 880, on March 31, 1959. This was the first to fly with sound suppressors on its CJ-805-3 turbojets. The fourth aircraft, flown on August 10, was the first of the Delta fleet and was also used in the FAA certification programme, which began in September.

Certification of the 880 was obtained on May 1, 1960. Delta Air Lines accepted delivery of its first aircraft on February 9, 1960, and the type went into service in May. Deliveries to TWA were delayed by financial difficulties and the Hughes Tool Co., which had ordered the aircraft as TWA's major shareholder, eventually leased six of the thirty on order to Northeast Airlines, which began service with the type on December 15, 1960. TWA put the first of twenty into service in January, 1961, and Hughes accepted delivery of the last four of the thirty-aircraft order in 1962. Delta increased its order to seventeen, with the final four for 1962 delivery, completing production of the basic 880.

CONVAIR 880 (U.S.A.)

A Convair 880 of the TWA fleet of twenty.

CONVAIR 880 (U.S.A.)

Convair 880-M. Developed as an intercontinental version of the Convair 880, as noted above, the 880-M or Model 22-M has three extra fuel tanks in the centre section of the wing. The increased fuel capacity led to higher gross weights, the final figure being 193,000 lb. To keep the airfield performance within reasonable limits, leading-edge flaps were introduced, as well as more powerful engines (11,650-lb. CJ-805-3Bs), a power-boosted rudder and strengthened undercarriage.

The first Convair 880 was modified to the new configuration and first flew on October 3, 1960. Orders for this version came from Avensa, which had become Viasa before the first of two was delivered in June, 1961; from Civil Air Transport (Air Asia) of Formosa, which received one in June, 1961; from Alaska Airlines, one received in July, 1961; and from Japan Air Lines, which received the first of five in July, 1961. One was also purchased by the Federal Aviation Agency and seven were built for a Capital Airlines order which was later cancelled. Two of these aircraft were used on lease by Swissair in 1961–62 and one sold to Cathay Pacific was delivered in February, 1962.

The FAA type approval for the Model 880-M was obtained on July 24, 1961.

CONVAIR 880-M

Dimensions

Span: 120 ft. 0 in.
Overall length: 129 ft. 4 in.
Overall height: 36 ft. 4 in.
Gross wing area: 2,000 sq. ft.
Sweepback: 35 deg. at 30% chord.
Internal cabin dimensions:
  Length (ex. flight deck): 89 ft. 3 in.
  Max. width: 129 in.; max. height: 85 in.
Max. usable floor area (ex. flight deck): 803 sq. ft.
Max. usable volume (ex. flight deck): 4,570 cu. ft.

Accommodation

Typical mixed class: 48F (4 abreast at 38-in. pitch) and 59E (5 abreast at 38-in. pitch).
Max. high density: 120 (5 abreast at 34-in. pitch).
Volume of freight and baggage holds: 849 cu. ft.

Powerplants

Four General Electric CJ-805-3B turbojets.
Take-off power each (I.S.A., s.l.): 11,650 lb. thrust at 7,684 r.p.m.
No water injection. Sound suppressors and thrust reversers fitted.

Weights and loadings

Basic operational: 93,100 lb.
Total fuel: 84,554 lb.
Max. (space limited) payload: 24,165 lb.
Max. take-off: 193,000 lb.
Max. landing: 155,000 lb.
Max. zero fuel: 121,500 lb.
Wing loading (max. t-o. wt.): 96.35 lb./sq. ft.
Wing loading (max. landing wt.): 77.50 lb./ sq. ft.
Power loading (max. t-o. wt.): 4.13 lb./lb. thrust.

Performance

Best cost. cruising speed (I.S.A.): 484 knots at 35,000 ft.; consumption: 9,450 lb./hr.
High-speed cruise: 534 knots at 22,500 ft.
Take-off field lengths (FAA requirements at max. t-o. wt.):
  At I.S.A. at sea level: 8,750 ft.
  At I.S.A.+15° C. at sea level: 9,850 ft.
Landing field length (FAA requirements at max. landing wt.): 5,930 ft.
Range (take-off to landing), still air, no reserves, I.S.A., max. fuel (10,436 imp. gal.): 4,050 n. mi. with 15,846-lb. payload at 439 knots (mean) at 35,000 ft. (mean).
Range (take-off to landing), still air, no reserves, I.S.A., with max. payload: 3,560 n. mi. at 438 knots (mean) at 35,000 ft. (mean).

Structure. Quasi-elliptical section fuselage with heavy-gauge 2,024 alloy skin. Three-spar wing with built-up spars and roll-tapered skins; 2,024-T4 lower skin, 7,075 upper skin.

Fuel system. Two integral tanks in each wing give basic 880 a capacity of 8,970 imp. gal. Model 880-M has centre-section tanks also for total capacity of 10,508 imp. gal. Four-point pressure refuelling at 1,000 i.g.p.m.

1. *The first Convair 880, N801TW, in Convair markings.*
2. *Delta Air Lines' Convair 880, N8802E.*
3. *One of the TWA 880, leased to Northeast as N8483H.*
4. *N801TW modified as the prototype 880-M with leading-edge flaps and other features.*

Undercarriage. Four-wheel main bogies, inwards retracting. Twin-wheel steerable nosewheel, forwards retracting. Track, 18 ft. 10½ in. Base, 52 ft. 11 in. Tyre pressure, 120–140 p.s.i.

Flying controls. Fully manual inboard ailerons with servo tabs and differentially operated inner and outer spoilers. Manual rudder (with power boost on 880-M) and manual elevator with servo tabs. Hydraulic variable-incidence tailplane with electric and manual back-up. Bendix PB-20G autopilot.

---

5. *The CAT Convair 880-M leaving San Diego with temporary U.S. registration N8486H.*
6. *Alaska Airline's 880-M, N8477H.*
7. *One of the two Viasa Convair 880-Ms, YV-C-VIA.*
8. *First of five 880-Ms for Japan Air Lines, JA8021.*
9. *One of the two 880-Ms leased by Swissair in 1961.*
10. *The Cathay Pacific Convair 880-M, VR-HFS.*

Flaps, air brakes. Double-slotted Fowler-type flaps in three pieces each side. Four-piece leading-edge flaps on each wing interconnected with flaps, on Model 880-M. Four upper-wing spoilers ahead of flaps, and u/c as air brake up to max. speed.

Cabin conditioning. Hamilton Standard Freon vapour-cycle system. Two turbocompressors driven by engine-bleed air. Differential, 8.3 p.s.i.

Hydraulics. Four engine-driven pumps for dual 3,000-p.s.i. system for flaps, u/c, spoilers, variable-incidence tailplane, brakes and nose-wheel steering; flaps and spoilers operate from either system.

Electrics. Four 40-kVA. engine-driven generators provide 115/208-V. 400-cycle a.c.; four transformer rectifiers for 28-V. d.c.; and 28-V. battery.

De-icing. Wing and intakes auto-iced by engine-bleed air; electric tail unit de-icing; Nesa glass screen fog removal and air-blast rain removal.

CONVAIR 880 (U.S.A.)

CONVAIR 990 (U.S.A.)

Photographed before delivery, this Swissair Convair 990 carries both Swiss (HB-ICA) and U.S. (N8497H) markings.

CONVAIR 990 (U.S.A.)

AS A GROWTH version of the Convair 880 (q.v.), the Convair 990 was projected early in 1958, when it was identified as the Model 30. Its new features comprised a lengthened fuselage; a larger wing with more fuel capacity; turbofan engines; and anti-shock fairings on the wing to allow a higher cruising speed to be reached for a given engine thrust setting. On July 30, 1958, the first airline contract was announced, when American Airlines ordered twenty-five.

The General Electric CJ-805-21 engines on which the design was based were derived from the -3 versions in the Convair 880, with the addition of a fan behind the turbine. Air ducted round the engine was compressed as it passed through the fan, to mix with the hot exhaust gases from the turbine. The efficiency of the engine was so improved in this way that the power was increased to 16,100 lb., while the specific fuel consumption was reduced by some 15 per cent.

To take advantage of the extra power, Convair refined the wing by increasing the chord without increasing the thickness. This gave a thinner effective wing section, the speed characteristics of which were further enhanced by the addition of the four anti-shock bodies on the trailing edge. These fairings were based on the work of Whitcomb at the NACA and were an application of area rule. They also provided additional space for fuel stowage; the capacity in the Model 30 wing was otherwise the same as that in the Convair 880-M (see p. 39).

To increase the payload, the fuselage was lengthened by 10 ft., allowing space for three more seat rows. The gross weight also went up, because of the higher structure weight, greater fuel load and larger payload. To keep airfield performance within reasonable limits, full-span leading-edge flaps were developed.

When first ordered, the Model 30 was publicly identified as the Convair 600, resurrecting an earlier title for the 880. As American Airlines felt that the lower number implied this was an older aeroplane than the Convair 880, however, the number was later changed to Convair 990. During 1959, Swissair ordered five Convair 600s (later increasing this to seven to allow two to be leased to SAS) and the name Coronado was bestowed on the type

by the Swiss airline. SAS itself ordered two Coronados in November, 1959, but had to cancel these early in 1962 through financial difficulties. The REAL Aerovias order for four Convair 880s was modified to three 990s, these aircraft being destined for operation by Varig in 1962.

Convair built no prototype of the 990, but put the type straight into production. The first to fly was therefore one of the American Airlines order (N5601), although airline markings were not carried. The first flight was made on January 24, 1961, from San Diego. Apart from the features already described, it had Krüger flaps under the wing leading edge between the fuselage and inner nacelles.

These flaps, adopted late in the development of the design, were a refinement of those first used on the Boeing 707 leading edge and comprised a flat plate hinged along the edge nearest the wing leading edge to open down into the airflow and improve elevator effectiveness by changing the pattern of the airflow.

Early in the flight trials, Convair discovered a serious snag in the outer engine installation. At certain combinations of airspeed and manœuvre (about 350 knots I.A.S. in the roll), an oscillation occurred in the outer pods which did not damp out sufficiently quickly of its own accord.

On March 21, 1961, therefore, N5601 was returned to the factory for modification. The change comprised a shortening of the outer pylon, which moved the engine position aft relative to the wing and resulted in a stiffer structure. Flight trials began again on April 20; meanwhile, the second 990 had begun its trials, without modification, on March 30.

The modified 990 reached a speed of Mach 0.97 at 22,500 ft. on May 8, 1961, equal to a true airspeed of 675 m.p.h. Further tests revealed, however, that the airframe drag was

1. *The first Convair 990, N5601, at take-off.*
2. *A later picture of N5601 after modification of the outer pylons.*
3. *Location of the anti-shock bodies on the wing of the Convair 990 is shown in this view.*
4. *Another picture of N5601, the first Convair 990.*

5

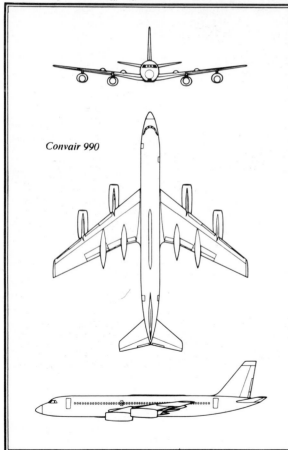

Convair 990

higher than expected and the guaranteed speed and range could not be achieved at maximum cruise thrust ratings of the engine. Exploring this problem delayed deliveries of the 990, and when these did begin, the aircraft were without a further series of modifications designed to restore the performance to its original level.

FAA certification of the 990 was obtained in mid-December, 1961, and American Airlines accepted delivery of its first aircraft on January 8. The first for Swissair was delivered on January 15 and arrived in Zurich on January 19. The first four Swissair Coronados

were quickly delivered and this airline became the first to put the type into scheduled service, at the end of February. The first American Airlines service, New York to Chicago, was flown on March 18, 1962, by which time the fifth Swissair aeroplane, in SAS colours, was ready for delivery.

Modifications for the Convair 990 were test-flown on the first aircraft early in 1962 and plans were made for retrospective modification of those in service later in 1962. These changes comprise: a fairing on the leading edge of the outer engine pylons; fairings on the inner face of all four pylons to slow down the airflow at the pylon-wing intersection; longer, less sharply tapered fairings over the engine thrust reversers; and a re-contoured leading edge with no slats but with Krüger flaps over the whole span.

5. *The fifth Convair 990, N5605, in American Airlines markings.*

CONVAIR 990

Dimensions

Span: 120 ft. 0 in.
Overall length: 134 ft. 9 in.
Overall height: 39 ft. 6 in.
Gross wing area: 2,250 sq. ft.
Sweepback: 35 deg. at 30% chord.
Internal cabin dimensions:
    Length (ex. flight deck): 98 ft. 9 in.
    Max. width: 129 in.; max. height: 85 in.
Max. usable floor area (ex. flight deck): 1,053 sq. ft.
Max. usable volume (ex. flight deck): 6,113 cu. ft.

Accommodation

Typical mixed class: 32F (4 abreast at 38-in. pitch) and 72E (5 abreast at 38-in. pitch).
Max. high density: 131 (5 abreast at 34-in. pitch).
Volume of freight and baggage holds: 928 cu. ft.

Powerplants

Four General Electric CJ805-23B turbofans.
Take-off power each (I.S.A., s.l.): 16,100 lb.s.t.
No water injection. No sound suppressors. Thrust reversers fitted.

Weights and loadings

Basic operational: 120,560 lb.
Total fuel: 105,000 lb.
Max. (space limited) payload: 26,440 lb.
Max. take-off: 246,200 lb.

Max. landing: 202,000 lb.
Max. zero fuel: 160,000 lb.
Wing loading (max. t-o. wt.): 109.5 lb./sq. ft.
Wing loading (max. landing wt.): 89.8 lb./sq. ft.
Power loading (max. t-o. wt.): 3.83 lb./lb.s.t.

Performance

Best cost. cruising speed (I.S.A.): 484 knots at 35,000 ft. and 190,000 lb.; consumption: 1,658 U.S. gal./hr.
High-speed cruise: 543 knots at 21,200 ft.
Long-range cruise: 420 knots at 35,000 ft.
Approach speed ($1.3V_{s0}$ at max. landing wt.): 157 knots.
Take-off field lengths (FAA requirements at max. t-o. wt.):
    At I.S.A. at sea level: 9,300 ft.
    At I.S.A. + 15° C. at sea level: 10,650 ft.
    At I.S.A. at 5,000 ft.: 13,100 ft.
Landing field length (FAA requirements at max. landing wt.): 6,550 ft.
Range (take-off to landing), still air, no reserves, I.S.A., max. fuel (13,050 imp. gal.): 4,730 n. mi. with 19,440-lb. payload at 450 knots (mean) at 35,000 ft. (mean).
Range (take-off to landing), still air, no reserves, I.S.A., with max. payload: 3,800 n. mi. at 450 knots (mean) at 35,000 ft. (mean).

Structure. Quasi-elliptical section fuselage with heavy-gauge 2,024 alloy skin. Three-spar wing with built-up spars and machine-tapered skins; 2,024-T4 lower skin, 7,075 upper skin.

Fuel system. Two integral tanks in each wing, plus centre-section tanks and anti-shock body tanks for total capacity of 13,049 imp. gal. Four-point pressure refuelling at 1,000 i.g.p.m.

Undercarriage. Four-wheel main bogies, inwards retracting. Twin-wheel steerable nosewheel, forwards retracting. Proportional anti-skid system.

Flying controls. Fully manual inboard ailerons with servo tabs and differentially operated inner and outer spoilers. Hydraulic-manual rudder with *q* feel. Manual elevator with servo tabs. Hydraulic variable-incidence tailplane with electric and manual back-up. Sperry S.P.30 autopilot.

Flaps, air brakes. Fowler-type flaps in three pieces each side. Krüger flaps under leading edge. Four upper-wing spoilers ahead of flaps, and u/c as air brake up to max. speed.

Cabin conditioning. Hamilton Standard Freon system. Two turbocompressors driven by engine-bleed air. Differential, 8.3 p.s.i.

Hydraulics. Four engine-driven pumps for dual 3,000-p.s.i. system for flaps, u/c, spoilers, variable-incidence tailplane, brakes and nose-wheel steering; flaps and spoilers operate from either system.

Electrics. Four 40-kVA. engine-driven generators provide 115/208-V. 400-cycle a.c.; four transformer-rectifiers for 28-V. d.c. and 28-V. battery.

Anti-icing. Wing and intakes anti-iced by engine-bleed air; electric tail unit de-icing; Nesa glass screen fog removal and air-blast rain removal.

DE HAVILLAND COMET (Great Britain)

ON MAY 2, 1952, a de Havilland Comet 1 left London Airport to inaugurate the world's first pure-jet airline service. Six and a half years later, a Comet 4 became the first jet airliner to operate over the North Atlantic route between London and New York. The operator of both these flights, which were among the most significant of any in the history of commercial airline flying, was BOAC. The two events were highlights in the chequered history of the Comet.

The history began early in 1943, when a committee which had been set up to consider the civil aircraft likely to be needed in Britain after the end of the war, included among its recommendations the development of a pure-jet mail carrier. This far-sighted recommendation was made at a time when only one British jet-propelled aeroplane had flown. The idea of a mail carrier was later abandoned but a second committee under the chairmanship of Lord Brabazon retained the idea of a pure-jet transport. In May, 1944, the requirement was considered to be an aeroplane able to carry 14 passengers and a small amount of mail for a distance of 700–800 miles.

The de Havilland company became interested in the possibilities of a civil jet transport before the end of 1943, and began making design studies of aircraft suitable to meet the Brabazon Committee recommendations.

In size, as well as layout, the aeroplanes being considered at this stage were far removed from the Comet. Maximum accommodation would be for twelve passengers for 1,400 miles, or only six passengers across the Atlantic. Although the Brabazon Committee had been thinking of much shorter ranges, both de Havilland and BOAC—the latter as potential operators—were anxious to make a transatlantic aeroplane out of the project. This inevitably led to considerable weight growth.

The general requirement was laid down in Specification 20/44 issued by the Ministry of Aircraft Production late in 1944, and in February, 1945, it was officially agreed that de Havilland should proceed with the design of an aeroplane to meet this specification. The type number D.H.106 was then allocated, and a tailless layout was chosen as the most promising for further development. It was to be powered by four de Havilland Ghost engines.

Both de Havilland and BOAC came to the conclusion that this tailless design was too great a risk and by the end of 1945 work switched to an alternative in which the wing sweep-back was reduced slightly and a swept-back tail unit was added. Contractual negotiations were based on this revised design and in May, 1946, a contract for two prototypes was placed with the MoS, with BOAC undertaking to buy about ten in due course.

The aeroplane as ordered was to have a gross weight of 93,000 lb., of which the payload would be 7,500 lb. The wing had 40 degrees of sweepback on the quarter-chord line and the span was 96 ft. 7 in. Four de Havilland Ghost engines were buried, so far as was possible, within the wing aft of the main spar.

Before the BOAC production order was confirmed, the D.H.106 again underwent a major revision by which the amount of sweepback on the wings was halved, to 20 degrees; the fuselage was lengthened to increase the basic accommodation from twenty-four to thirty-two; and an unswept tail unit was adopted. These changes increased the wing span to 111 ft. and the gross weight to 100,000 lb., whilst reducing the max. cruising speed from 535 m.p.h. to 505 m.p.h. The engines remained the same.

This was the D.H.106 as adopted formally by BOAC on January 21, 1947, when a preliminary order for eight aircraft was placed. This was later increased to nine after BOAC had absorbed British South American Airways, which had ordered six Comets subsequent to the first BOAC order. At the time of this first order from BOAC, no other airline in the world had contracted to purchase a turbine engine transport of any type, either turbojet or turboprop.

The first of the two prototypes, carrying the markings G-5-1, was rolled out on July 25, 1949, and taxy trials were made on July 26 and July 27. Towards the evening of the latter day, John Cunningham flew the Comet off for its first flight.

For the next year, the first prototype (G-ALVG) was flown intensively. It was joined exactly one year later, on July 27, 1950,

DE HAVILLAND **COMET** (Great Britain)

One of the BOAC fleet of de Havilland Comet 4s over London.

DE HAVILLAND COMET (Great Britain)

by G-ALZK, the second prototype. Both these aircraft were fitted with the original single wheel main undercarriage units, although G-ALVG was used in December, 1950, for a trial fixed installation of the production-type four-wheel bogie. On May 7, 1951, it began a new series of flights in which two Sprite rocket motors in the wings were used to assist the take-off. Structural provisions had been made in the wings for such an installation, in the belief that power boosting might be needed by some Comet operators. Both prototypes were used in due course to test later Comet features—a droop-snoot leading edge on G-ALVG in December, 1952, and the wing "pinion" tanks on G-ALZK in 1952.

Production of the Comet has been shared between the de Havilland factories at Hatfield and Chester and has been in six principal versions. Prototypes, experimental versions and projects have accounted for nine or ten other designations in the Comet series, all of which are included in the notes which follow.

Comet 1. The nine aircraft for BOAC were built to this initial production standard, similar in appearance to the prototypes apart from the bogie undercarriage. The span was 115 ft., length 93 ft. and gross weight 105,000 lb., later increased to 107,000 lb. Four 5,050-lb.s.t. Ghost 50 Mk.1 engines were fitted and the fuel capacity was 6,050 imp. gal.

The first production aircraft (G-ALYP) flew on January 9, 1951, followed by three more during the year. A normal category certificate of airworthiness—the first ever awarded for a jet aeroplane—was issued on January 22, 1952, and BOAC operated the first jet service with G-ALYP between May 2 and May 6, on the London-Johannesburg-London route. The first Comet service on the eastern route to Colombo, Ceylon, was flown on August 12, 1952, by G-ALYU.

Four of the nine Comet 1s were destroyed in accidents between October, 1952, and January, 1954. The first, at Rome on October 26, was attributable to the take-off technique and lift-characteristics of the wing and was in part responsible for development of the drooped leading edge already mentioned. The other three—Calcutta, May 2, 1953; Elba, January 10, 1954, and Naples,

April 8, 1954—showed a pattern of similarity of sudden en route disaster and resulted in the grounding of the Comet fleet in April, 1954, and the most exhaustive of accident investigations ever undertaken. This concluded that the Comets had suffered structural failure of the fuselage originating with a fatigue crack near a cut-out in the skin.

Comet 1A. After BOAC had taken the initiative in ordering the Comet, several other operators were quick to follow suit. Orders came from Canadian Pacific, for two in December, 1949; from the French operator UAT, for three in May, 1951; from the RCAF for two and from Air France for three in November, 1951. All these were built as Comet 1As, the chief difference from the Comet 1 being an increase in fuel capacity to 6,906 imp. gal. and in gross weight to 110,000 lb. (later 115,000 lb.).

The first to fly was CF-CUM, for CPA, on August 11, 1952. This airline did not put the Comet into service, however, as one of its two aircraft crashed on its delivery flight. The first Comet 1A service, therefore, was on February 19, 1953, by UAT between Casablanca and Dakar. Air France started operating Comets on August 26, 1953, on the route Paris-Rome-Beirut, and the RCAF took delivery of their two in May and June, 1953.

All Comet 1As were grounded in April, 1954, together with the 1s.

Comet 1XB. After the reason for the Comet 1 accidents had been established, some of the Comet 1As were modified for further limited-life service in military guise. The aircraft concerned were the two RCAF Comets, which returned to service in September, 1957, and the three Air France aircraft which were purchased by the British Ministry of Supply and allocated for special duties in the U.K.

---

1. *The first prototype Comet, G-ALVG, with the original single-wheel undercarriage.*
2. *Second prototype Comet, G-ALZK.*
3. *A production Comet 1, G-ALYP, in BOAC colours.*
4. *The Comet 2X, G-ALYT, with Avon engines.*

Comet 2X. Possible development of the Comet 1 design by using more powerful engines was discussed by de Havilland's and BOAC before the end of 1946, and orders for a Series 2 version with Rolls-Royce Avon engines followed in due course. The Ministry of Supply purchased one production Series 1 airframe for Avon development, under the designation Comet 2X. With 6,500-lb.s.t. Avon RA.9 Mk. 501 (later Mk. 502) engines, it first flew on February 16, 1952. In 1957, a thrust reverser was fitted on one engine, and a water-spray rig was also fitted for engine de-icing trials.

Comet 2. The second production series of Comets had 7,300-lb. Avon RA.25 (Mk. 503 and Mk. 504) engines, a three-foot-longer fuselage and gross weight of 120,000 lb. The fuel capacity remained as in the Comet 1A.

BOAC ordered twelve of the new Comets, and other orders came from CPA (three), UAT (three), Air France (six), BCPA (three), Japan Air Lines (two), LAV (two) and Panair do Brasil (four). Production lines were established at Hatfield, Chester and Belfast (Short Bros. and Harland), and the first aircraft at Hatfield (G-AMXA) flew on August 27, 1953. Others followed, and twenty-two were in various stages of flight test or production when the Comet 1s were grounded and all work on the Series 2 was brought to a halt.

Of these twenty-two, ten eventually went into service with RAF Transport Command and three others with No. 90 Group; two were used for engine development, one was used for pressure tests, three were put into long-term storage and parts for the other five were scrapped.

Comet 2E. Two of the original BOAC order for Comet 2s, used for Avon RA.29 development for the Comet 4. Each had an Avon RA.29

---

5. *An RCAF Comet after modification to 1XB standard.*
6. *Air France Comet 1A, F-BGNX.*
7. *The CPA Comet 1A, CF-CUM.*
8. *First of three UAT Comet 1As, F-BGSA.*

(Mk. 524) in each outer position and an RA.25 (Mk. 504) inboard. The first (G-AMXD) flew in April, 1957, after modification. They were later allocated for radio and radar development.

Comet T.2. The first two of ten Comet 2s modified for use by No. 216 Squadron, RAF Transport Command, were fitted out as crew trainers and designated T.2. The first (XK669) flew after modification on December 9, 1955.

Comet C.2. Eight Comet 2s laid down for commercial operators but completed for No. 216 Squadron, RAF Transport Command with modifications to permit an 8,000-hr. life.

Comet 2R. Three of the early production Comet 2s for BOAC were modified for use by No. 90 Group (later Signals Command), RAF, in radar and electronic development, and carry this unofficial designation.

Comet 3. In the autumn of 1952, de Havilland announced a new version of the Comet with a fuselage lengthened to 111 ft. 6 in., fuel capacity up to 8,308 imp. gal. by use of "pinion" tanks on the wings and weight increased to 145,000 lb. A prototype was ordered by the Ministry of Supply; BOAC confirmed an order for five in 1954, and orders were also obtained from Pan American (for three) and Air-India (two).

Despite the grounding of Comet 1s, work on the prototype (G-ANLO) of the Comet 3 continued and the first flight was made on July 19, 1954, with Avon RA.26s (Mk. 522). No production took place, the similar Comet 4 being developed instead. On February 25, 1957, G-ANLO began a new series of trials after modification to incorporate a number of features of the Series 4, and Avon RA.29 (Mk. 523) engines.

Comet 3B. After BEA had ordered the Comet 4B in 1957, the Comet 3 G-ANLO was modified to have the new short-span wing without pinion tanks, and outer-engine thrust reversers. The first flight in this guise was on August 21, 1958. During 1961, the Comet 3B was allocated to the Blind Landing Experimental Unit at RAE Bedford.

DE HAVILLAND COMET (Great Britain)

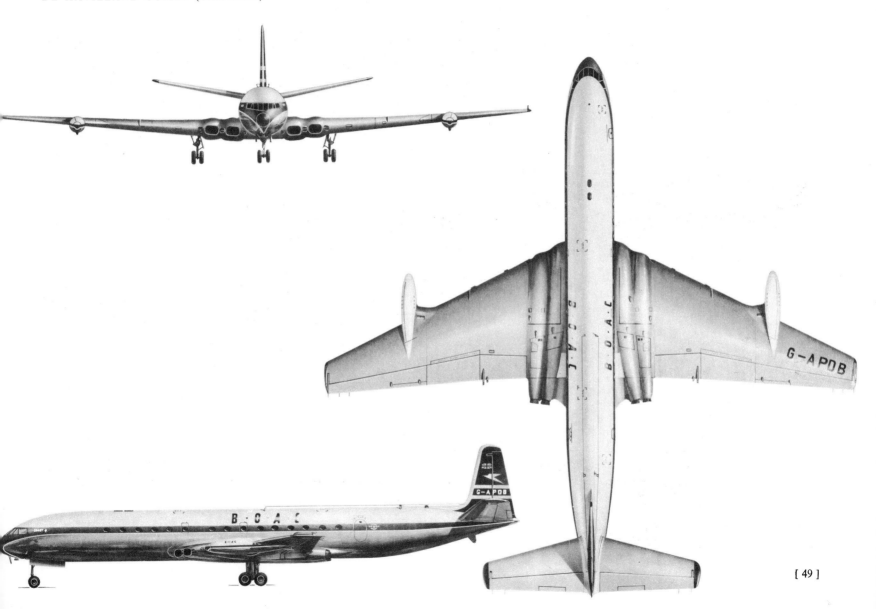

Comet 4. Immediately after the Court of Inquiry into the Comet 1 accidents had produced its report, BOAC contracted to buy a fleet of nineteen Comets similar to the Series 3 but with enough modifications to justify a new designation. While production lines were put down at Hatfield and Chester, the Comet 3 completed about 80 per cent. of the certification trials, while the two Comet 2Es built up engine hours. Consequently, the Comet 4 was able to go into service only five months after the first flight on April 27, 1958, and on October 4 in that year BOAC operated the world's first jet airliners across the North Atlantic in each direction between London and New York.

Since then, BOAC Comet 4s have been withdrawn from that route for use on the Eastern and Southern services to Australia, South Africa and the Far East, for which they were originally intended.

Towards the end of 1959, Qantas began chartering Comet services from BOAC for flights between London and Singapore, and similar arrangements were made by Nigeria Airways and Air Ceylon in March, 1962, and Air India in April, 1962. Aerolineas Argentinas put the first of six Comet 4s into service in May, 1958, and East African Airways has three, first in service in September, 1960.

Comet 4B. After a Comet 4A had been projected for Capital Airlines with the wing cropped by 7 ft. and the fuselage lengthened by 40 in., de Havilland increased the fuselage length still further, to 118 ft., in the Comet 4B. The clipped wing of 108-ft. span was retained but the pinion tanks were deleted. The fuel capacity then went down to 7,890 imp. gal. while the gross weight went up to 156,000 lb. BEA ordered six of this short-range version, with 10,500-lb.s.t. Avon RA.29 (Mk. 525) engines with thrust reversers and sound suppressors, and later increased this to fourteen. Olympic Airways bought four. The first Comet 4B flew on June 27, 1959, and went into service with BEA in April, 1960.

Comet 4C. Final Comet version, the 4C combined the long fuselage of the 4B with the full-span wing and fuel capacity of the Series 4. Orders for this version have come from Mexicana (three), United Arab Airlines (seven), Middle East Airlines (four), Aerolineas Argentinas (one), King Saud of Saudi Arabia. (one) and the RAF (five, designated C.Mk.4C). The first Series 4C, one of the Mexicana order, flew on October 31, 1959. Two of the Mexicana 4Cs operated for a time on charter to Guest Aerovias, another Mexican airline.

DE HAVILLAND **COMET 4**

Dimensions

Span: 115 ft. 0 in. (4B, 107 ft. 10 in.)
Overall length: 111 ft. 6 in. (4B and 4C, 118 ft.)
Overall height: 29 ft. 6 in.
Gross wing area: 2,121 sq. ft. (4B, 2,059 sq. ft.)
Sweepback: 20 deg. on quarter chord.
Internal cabin dimensions:
    Length (ex. flight deck): 71 ft. 8 in. (4B and 4C, 78 ft. 2 in.)

    Max. width: 115 in.; max. height: 78.5 in.
Max. usable floor area (ex. flight deck): 439 sq. ft. (4B and 4C, 529 sq. ft.)
Max. usable volume (ex. flight deck): 2,815 cu. ft. (4B and 4C, 3,160 cu. ft.)

Accommodation

Typical mixed class (4C): 20F (4 abreast at 40-in. pitch) and 59E (5 abreast at 34-in. pitch).
Max. high density: 81 (4B, 4C, 102) (5 abreast at 34-in. pitch).
Volume of freight and baggage holds: 570 cu. ft.

Powerplants

Four Rolls-Royce Avon 524 turbojets (4B, Avon 525B).
Take-off power each (I.S.A., s.l.): 10,500-lb. thrust.

9. *A Comet 2, G-AMXD, in BOAC markings.*
10. *First of the ten RAF Comet 2s, the T.2 XK669.*
11. *The Comet 3 prototype G-ANLO.*
12. *G-ANLO after conversion to Comet 3B.*

DE HAVILLAND **COMET** (Great Britain)

Sound suppressors fitted (4B, thrust reversers, outer engines).

COMET 4

Weights and loadings

Basic operational: 75,424 lb.
Total fuel: 71,220 lb.
Mfrs. max. payload: 20,286 lb.
Max. take-off: 162,000 lb.
Max. landing: 120,000 lb.
Max. zero fuel: 99,000 lb.
Wing loading (max. t-o. wt.): 76.4 lb./sq. ft.
Wing loading (max. landing wt.): 56.6 lb./sq. ft.
Power loading (max. t-o. wt.): 3.86 lb./lb. thrust.

Performance

Best cost cruising speed (I.S.A.): 438 knots at 28,000 ft. and 130,000 lb.; consumption: 9,650 lb./hr.
Approach speed ($1.3 V_{s_0}$ at max. landing wt.): 124 knots.
Take-off field lengths (BCAR requirements at max. t-o. wt.):
At I.S.A. at sea level: 6,750 ft.
At I.S.A. +15° C. at sea level: 7,500 ft.
At I.S.A. at 5,000 ft.: 9,300 ft.
Landing-field length (BCAR requirements at max. landing wt.): 6,630 ft.
Range (take-off to landing), still air, no reserves, I.S.A., max. fuel (8,898 imp. gal.): 3,820 n. mi. with 15,476-lb. payload at 405 knots (mean) at 37,000 ft. (mean).
Range (take-off to landing), still air, no reserves, I.S.A., with max. payload: 3,500 n. mi.

COMET 4B

Weights and loadings

Basic operational: 78,363 lb.
Total fuel: 62,400 lb.
Mfrs. max. payload: 24,137 lb.
Max. take-off: 158,000 lb.
Max. landing: 120,000 lb.
Max. zero fuel: 102,500 lb.
Wing loading (max. t-o. wt.): 76.7 lb./sq. ft.
Wing loading (max. landing wt.): 58.2 lb./sq. ft.
Power loading (max. t-o. wt.): 3.76 lb./lb. thrust.

Performance

Best cost. cruising speed (I.S.A.): 452 knots at 23,500 ft. and 135,000 lb.; consumption: 11,500 lb./hr.
Approach speed ($1.3 V_{s_0}$ at max. landing wt.): 128 knots.
Take-off field lengths (BCAR requirements at max. t-o. wt.):
At I.S.A. at sea level: 6,600 ft.
At I.S.A. +15° C. at sea level: 7,450 ft.
At I.S.A. at 5,000 ft.: 9,300 ft.

13. *Aerolineas Argentinas Comet 4, LV-PLM.*
14. *The Comet 4 VP-KPJ for East African Airways.*
15. *A BEA Comet 4B, G-APMA.*
16. *Olympic Airways Comet 4B with British registration G-APYC.*
17. *A BOAC Comet 4, G-APDP, on lease to Quantas.*

13

14

15

16

17

Landing field length (BCAR requirements at max. landing wt.): 6,600 ft.

Range (take-off to landing), still air, no reserves, I.S.A., max. fuel (7,813 imp. gal.): 3,240 n. mi. with 17,131-lb. payload at 405 knots (mean) at 37,000 ft. (mean).

Range (take-off to landing), still air, no reserves, I.S.A., with max. payload: 2,910 n. mi.

COMET 4C

Weights and loadings—As Comet 4 except:
Basic operational: 77,500 lb.
Max. zero fuel: 102,500 lb.

Performance

Best cost. cruising speed (I.S.A.): 451 knots at 31,000 ft. and 130,000 lb.; consumption: 1,225 imp. gal./hr.

High-speed cruise: 451 knots at 31,000 ft.

Long-range cruise: 400 knots at 36,000 ft.

Approach speed ($1.3 V_{s0}$ at max. landing wt.): 126 knots.

Take-off field lengths (BCAR requirements at max. t-o. wt.):
At I.S.A. at sea level: 6,600 ft.
At I.S.A.$+15°$ C. at sea level: 7,450 ft.
At I.S.A. at 5,000 ft.: 9,150 ft.

Landing field length (BCAR requirements at max. landing wt.): 6,630 ft.

Range (take-off to landing), still air, no reserves, I.S.A., max. fuel (8,898 imp. gal.): 3,745 n. mi. with 11,445-lb. payload at 405 knots (mean) at 37,000 ft. (mean).

Range (take-off to landing), still air, no reserves, I.S.A., with max. payload: 2,910 n. mi.

Structure. Semi-monocoque circular-section fuselage. Two-spar stub wings and centre-section; semi-monocoque extension wings.

Fuel system. Integral tanks in stub and extension wings, bag tanks in stub and centre section. Two nacelle tanks on wings. Two-point pressure refuelling at 400 g.p.m.

Undercarriage. Four-wheel bogie main units and twin-wheel steerable nosewheel. Track, 28 ft. 2 in. Base, 53 ft. 2 in. Tyre pressure, 150 p.s.i.

Flying controls. All surfaces power operated by duplicated hydraulic and mechanical units. Ailerons have spring feel only; elevators spring feel plus q feel; and rudder, spring feel q restrictor. Smiths S.E.P.2 autopilot.

Flaps, air brakes. Plain outer and split-type inner flaps on each wing. Slotted plate air brakes above and below each wing.

Cabin conditioning. D.H. Propellers system, using engine-bleed air. Differential, 8.75 p.s.i.

Hydraulics. Four-circuit (plus handpump) 2,500-p.s.i. system with four engine-driven and two electric pumps for flying controls, u/c, flaps, air brakes, wheel brakes and nosewheel steering.

Electrics. Four 14-kVA. alternator/rectifier combinations feed busbar, normally single but divisible under fault conditions.

De-icing. Wings, tailplane, fin and intakes by hot air from engine compressors of all engines.

18. *XA-NAS, the second Comet 4C or Mexicana.*
19. *The first Misrair Comet 4C, SU-ALC.*
20. *After Misrair linked with Syrian Airways the name changed to United Arab Airlines, in whose markings SU-ALE is shown.*
21. *Middle East Airlines' Comet 4C, OD-ADR.*

DE HAVILLAND **TRIDENT** (Great Britain)

The de Havilland Trident G-ARPA on the occasion of its first flight.

FIRST FLOWN on January 9, 1962, the Trident is the first of a new generation of short-haul jet airliners, incorporating the lessons of the first five years of regular jet operations. Its origin springs from a requirement of July, 1956, drawn up by BEA, for a jet aeroplane suitable for its routes which were mostly below 1,000 miles in length.

The first de Havilland project to meet this specification was the D.H.119, using four Rolls-Royce Avon R.A.29s. For a short time, an attempt was made to combine the BEA requirement with one from BOAC for a much longer-range aeroplane. This "compromise" aeroplane, the D.H.120, proved impracticable, and after BOAC had ordered the Vickers VC10, de Havilland again concentrated on the BEA type, with their D.H.121 design.

A long period of political argument followed, while the Government sought to force BEA to purchase the project of what was considered the most satisfactory company. Eventually, a consortium of de Havilland, Hunting and Fairey was formed to produce the D.H.121

and in February, 1958, BEA was allowed to order this aircraft, which was considered to be technically the best of the projects considered.

The layout adopted by de Havilland used three engines (12,000-lb. Rolls-Royce R.B.141s initially) in the rear fuselage—similar arrangements were made in the Bristol 200 and Avro 740, which had been the principal contenders for the BEA order. For a year, work on the D.H.121 continued, but a major redesign then became necessary to reduce the size and weight to keep within the latest BEA requirements. Rolls-Royce Spey 10,000-lb.s.t. turbofans were then adopted, the gross weight was limited to 105,000 lb. and the BEA order for twenty-four was confirmed in August, 1959. When de Havilland became part of the Hawker Siddeley Group later in 1959, the Airco consortium was dissolved.

For BEA service, the Trident 1 will be delivered in the second half of 1963. A second version, the Trident 1C, has additional fuel in the centre section and a gross weight of 112,000 lb. No prototype has been built, the first to fly being the first for BEA, whose fleet of twenty-four will carry the registrations G-ARPA to G-ARPP, G-ARPR to G-ARPU and G-ARPW to G-ARPZ inclusive (constructors' numbers, 2101 to 2124 inclusive).

ABOVE—*The three engines grouped in the tail are a distinctive feature of the Trident.*

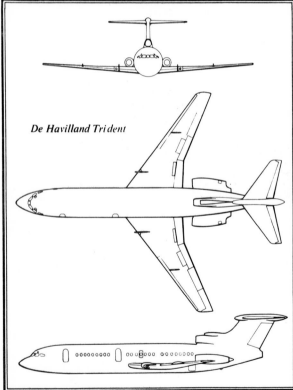

De Havilland Trident

DE HAVILLAND **TRIDENT 1**

Dimensions

Span: 89 ft. 10 in.
Overall length: 114 ft. 9 in.
Overall height: 27 ft. 6 in.
Gross wing area: 1,358 sq. ft.
Sweepback: 35 deg. on quarter-chord line.
Internal cabin dimensions:
    Length (ex. flight deck): 65 ft. 6 in.
    Max. width: 135.5 in.; max. height: 80 in.
Max. usable floor area (ex. flight deck): 708 sq. ft.
Max. usable volume (ex. flight deck): 4,440 cu. ft.

Accommodation

Typical mixed class: 25F (4 abreast at 40-in. pitch) and 58E (6 abreast at 34-in. pitch).
Max. high density: 104 (6 abreast at 34-in. pitch).
Volume of freight and baggage holds: 620 cu. ft.

Powerplants

Three Rolls-Royce RB163-1 Spey turbofans.

Take-off power each (I.S.A., s.l.): 10,400 lb. at 12,490 r.p.m.
No water injection. Sound suppressors and thrust reversers (outer engines only) fitted.

Weights and loadings

Basic operational: 63,130 lb.
Total fuel: 31,720 lb.
Mfrs. max. payload: 22,370 lb.
Max. take-off: 107,000 lb.
Max. landing: 100,000 lb.
Max. zero fuel: 85,500 lb.
Wing loading (max. t-o. wt.): 78.7 lb./sq. ft.
Wing loading (max. landing wt.): 73.6 lb./sq. ft.
Power loading (max. t-o. wt.): 3.4 lb./lb. thrust.

Performance

Best cost. cruising speed (I.S.A.): 508 knots at 32,000 ft. and 100,000 lb.; consumption: 914 imp. gal./hr.

High-speed cruise: 526 knots at 25,000 ft.
Long-range cruise: 461 knots at 35,000 ft.
Approach speed ($1.3 V_{S_0}$ at max. landing wt.): 134 knots.
Take-off field lengths (BCAR requirements at max. t.-o. wt.):
At I.S.A. at sea level: 5,800 ft.
At I.S.A. $+15°$ C. at sea level: 6,700 ft.
At I.S.A. at 5,000 ft.: 8,100 ft.
Landing field length (BCAR requirements at max. landing wt.): 6,370 ft.
Range (take-off to landing), still air, no reserves, I.S.A., max. fuel (3,840 imp. gal.): 2,350 n. mi. with 12,000-lb. payload at 461 knots (mean) at 35,000 ft. (mean).
Range (take-off to landing), still air, no reserves, I.S.A., with max. payload: 1,530 n. mi.

Structure. Circular-section monocoque fuselage. Two-spar wing with third spar over centre section and inner wings. Naturally aged Al-Cu alloys for skins and stringers.

Fuel system. Two-bay integral tank in each wing, Standard capacity 3,840 imp. gal.; max. capacity with optional centre-section tank. 4,840 imp. gal. (Trident 1C).

Undercarriage. Four-abreast main wheel units pivot through 90 deg. and fold sideways. Offset nosewheel folds sideways. Track, 18 ft. 8 in. Base, 43 ft. 6 in.

Flying controls. Two-section ailerons, inners only used at high speeds. All moving tailplane, geared elevator. Triplicated hydraulic actuators close to all controls. Smiths-Kelvin/Sperry integrated flight control system.

Flaps, air brakes. Double-slotted flaps in four sections with interconnected drooping leading edge. Lift dumpers forward of inner flaps, spoilers ahead of outer flaps. Undercarriage and reverse thrust used as air brakes.

Cabin conditioning. D.H. air-conditioning system fed by engine-bleed air. Provision for A.P.U. for ground use. Cabin pressure differential, 8.25 p.s.i.

Hydraulics. Three independent 3,000-p.s.i. systems powered by three engine-driven pumps, with independent stand-by, to operate flying controls, u/c, brakes, flaps, droop leading edge, slats, air brakes and lift dumpers.

Electrics. Three a.c. generators for 400-cycle three-phase 115/200-V. a.c. and, through three rectifiers, 28-V. d.c.

De-icing. Airframe de-icing, including engine and air intakes, by engine-bleed air. Electric film screen de-icing.

DOUGLAS **DC-8** U.S.A.

WITH A WORLD-WIDE reputation as a producer of commercial aeroplanes, from its pre-war DC-1, DC-2, DC-3 and DC-4, to the post-war DC-6 and DC-7 series, the Douglas Aircraft Company was the second U.S. manufacturer to produce a jet transport. When the search began, in 1952, for an aeroplane to succeed the DC-6/7 series, Douglas designers were led inexorably toward a jet-powered type, although turboprop projects were actively considered for some time.

By the time Douglas plans had become firm, the Boeing company already had decided to go ahead with construction of a prototype, aimed both at the commercial market and the USAF's requirement for a jet tanker. Although Douglas made no attempt to meet the military requirement, the design of the DC-8 (Douglas Commercial Eight) was remarkably similar to the Boeing 707 in overall appearance.

In the first instance, the DC-8 was designed to suit U.S. domestic operations, with sufficient range to fly across the continent non-stop. This produced an aeroplane of 211,000-lb. gross weight, powered by four Pratt & Whitney J57 engines; the span was to be 134.6 ft. and the length 140.7 ft. Higher weights were projected for overwater versions, and when the design was first publicly described in August, 1955, the highest weight mentioned was 257,000 lb. On October 13, 1955, Pan American Airlines became the first to order the DC-8, with a contract for twenty.

Once the decision had been taken (on June 7, 1955) to proceed with construction of the DC-8, Douglas designers found themselves in fierce competition with their Boeing opposite numbers to produce the aeroplane best able to meet the needs of the largest possible number of the world's airlines. Completely opposite policies were adopted by the two companies. While Boeing diversified their design into a whole family of variants, Douglas adhered rigidly to a single airframe design, varying only the powerplant and fuel capacity.

To match the competition, the DC-8 grew in size during 1956. More range and more payload were built in, forcing up the weight, with matching increases in span and length to 139.7 ft. and 150.5 ft. respectively. The gross weight of the domestic model went up to 265,000 lb. and of the overwater DC-8 to 287,500 lb., using the more powerful JT4A engines.

While detail design and component sub-assembly for the DC-8 continued during 1956, a steady flow of airline orders was received and a third basic variant was added to the range—an overwater or intercontinental type with Rolls-Royce Conway engines.

Although Series numbers were not adopted for the DC-8 variants until the end of 1959, the development of the type can conveniently be recorded from this point under the different Series headings.

DC-8 Series 10. This designation was adopted for the domestic model, with the smallest fuel capacity (17,600 U.S. gallons) and lowest gross weight (273,000 lb.) of any DC-8 variant. United Air Lines, which was the second customer for the DC-8 with an order announced on October 25, 1955, received twenty-two of this model, starting on June 3, 1959, with JT3C-6 engines. The only other customer for this version was Delta Air Lines, which had six. Both airlines inaugurated DC-8 service on September 18, 1959.

The first DC-8 to fly was built as a Douglas demonstrator, in United configuration with JT3C-6 engines, and this first flew from Long Beach on May 30, 1958. Bearing the Douglas registration N8008D, this particular aircraft served as a demonstrator until late in 1961, when it was delivered to National Air Lines after conversion to a Series 50. For the early flight trials it had air brakes on the fuselage sides just behind the wing fillets.

Other DC-8s used in the certification programme included two more Series 10s, both flown in December, 1958, and registered N8028D and N8038D prior to their delivery to United Air Lines. N8028D was the first to fly with the Douglas-designed sound suppressor/thrust reverser engine nacelles.

In the course of flight development of the first DC-8s of the Series 10, 20 and 30 type, several modifications were introduced. Most important of these were the slots in the wing leading edge inboard of each engine pylon, to improve the low-speed characteristics of the wing. The slots are normally closed by flush

DOUGLAS **DC-8** (U.S.A.)

A DC-8 Series 30 of TAI, one of two French independent airlines which purchased this type.

doors in upper and lower surfaces, which open automatically when the flaps are lowered. They were first flown on the second DC-8 and were introduced at the thirtieth production line position.

Sub-variants of the Series 10 are designated DC-8-11 and DC-8-12. The -11 was granted FAA type approval on August 31, 1959, at a gross weight of 265,000 lb. The -12 designation applies when wing leading edge slots and extended wing tips are fitted. This version was approved at a gross weight of 273,000 lb. on July 1, 1960. Fuel capacity of both variants was 119,244 lb. in eight wing tanks.

DC-8 Series 20. Several U.S. domestic operators took advantage of engine growth to order domestic models of the DC-8 with JT4A turbojets. The Series 20 had the same fuel capacity as the Series 10, but the gross weight went up to 276,000 lb. and the increased power improved the take-off performance.

The second DC-8 to fly was a Series 20, the date being November 29, 1958. This was registered N8018D and later went to United Air Lines as one of fifteen of this Series used by the airline. Eastern Airlines ordered Series 20s in December, 1955, and eventually received fifteen, relinquishing one other to Aeronaves de Mexico. National Air Lines is the only other Series 20 operator, with three ordered in November, 1955. The 15,800-lb. thrust JT4A-3 or -5 engines were first used in this variant, which received its FAA type certificate on January 19, 1960. The more powerful JT4A-9, -10, -11 or -12 can be substituted.

DC-8 Series 30. First of the intercontinental versions, the Series 30 was first ordered by Pan American at a gross weight of 287,500 lb., with JT4A-3 engines. The first example to fly was one of the Pan American fleet, on February 21, 1959, and two more joined the test programme soon after.

In June, 1958, Douglas had announced further development of the intercontinental, with additional fuel in the wing centre section, higher gross weight (310,000 lb.) and 16,800-lb. thrust JT4A-9 or -10 engines. Most airlines purchasing the Series 30 specified this configuration, but flight testing revealed a deficiency in performance of this version, with the maximum range 8–10 per cent. below the guarantee.

This was tracked down to excessive drag when the aircraft was being flown at the angle of attack for maximum range. The deficiency was cured by a series of modifications, the most important of which were low drag wing tips, extending the span to 142 ft. 5 in., and a new wing leading edge to alter the wing profile.

Modifications were developed on the seventh test DC-8, N8068D, a Pan American Series 30, starting in October, 1959. The new wing tips came in on the thirtieth aircraft but the modification which increased the wing chord by 4 per cent. and the gross area from 2,771 sq. ft. to 2,868 sq. ft., did not become standard until the one hundred and forty-eighth aircraft. It was then applied retrospectively to a number of DC-8s already in service.

Sub-variants of the Series 30 are designated DC-8-31, -32 and -33. The DC-8-31 was approved on March 30, 1960, at a gross weight of 300,000 lb. and the DC-8-32 was approved on February 1, 1960, at a gross weight of 310,000 lb., respectively without and with the extended wing tips. The DC-8-33 was approved on November 28, 1960, at a gross weight of 315,000 lb. with 17,500 lb. JT4A-11 or -12 engines. The fuel capacity was either 150,000 lb. with one additional tank in the centre section, or 157,000 lb. with two extra tanks. With the extended leading edge, the fuel capacity increases to 159,350 lb.

Of twenty-five DC-8 Series 30 ordered by Pan American, two were allocated to Panair de Brasil and four to Pan American Grace. Other Series 30s have been purchased by Northwest Orient Airlines (five), SAS (seven), KLM (seven), Swissair (three), Japan Air Lines (five), UAT (two) and TAI (three). Versions of the Pratt & Whitney engine fitted are the JT4A-3, -5, -9, -10, -11 and -12.

DC-8 Series 40. The Rolls-Royce Conway was first offered as an alternative to the JT4A-

---

1. *The first DC-8, N8008D, on an early test flight.*
2. *First DC-8 with JT4A engines, N8018D, on its first take-off.*
3. *A United Air Lines DC-8 Series 10, N8005U.*
4. *Delta Air Lines' first DC-8 Series 10, N801E.*

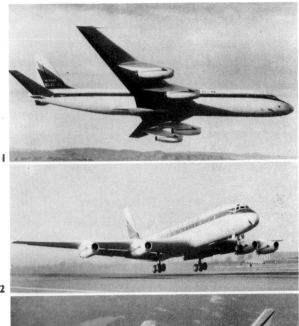

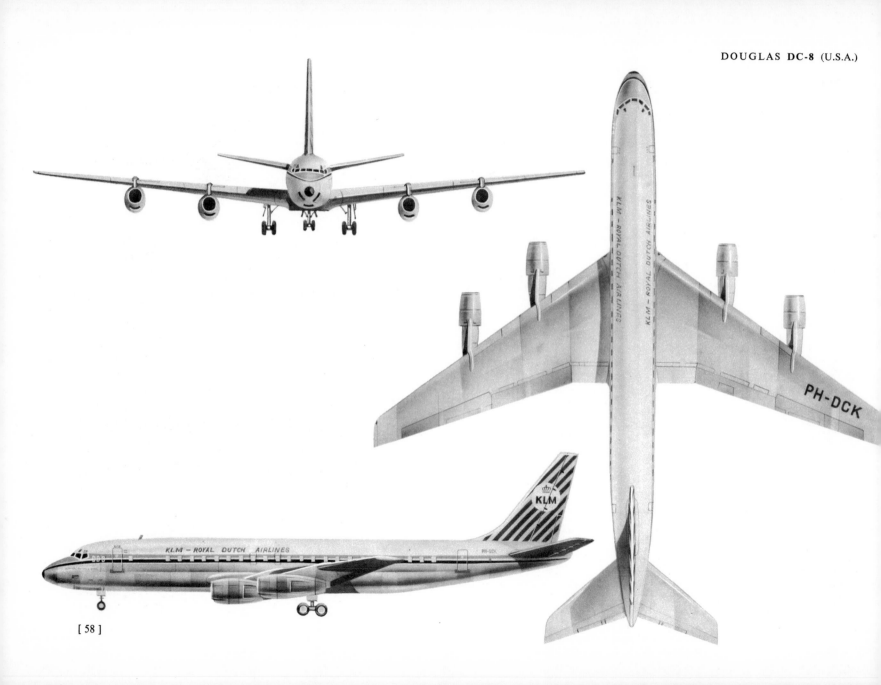

DOUGLAS **DC-8** (U.S.A.)

[58]

DOUGLAS DC-8 (U.S.A.)

DOUGLAS DC-8

Dimensions

Span : 142 ft. 5 in.
Overall length : 150 ft. 6 in.
Overall height : 42 ft. 4 in.
Gross wing area : 2,725 sq. ft.
Sweepback : 30 deg. at quarter-chord line.
Internal cabin dimensions :
  Length (ex. flight deck) : 103 ft. 0 in.
  Max. width : 140 in. ; max. height : 80 in.
Max. usable floor area (ex. flight deck) : 1,100
  sq. ft.
Max. usable volume (ex. flight deck) : 7,940 cu. ft.

Accommodation

Typical mixed class : 24F (5 abreast) and 93E
  (6 abreast).
Max. high density : 177 (6 abreast at 32-in. pitch).
Volume of freight and baggage holds : 1,390 cu. ft.

DC-8 Series 10

Powerplants

Four Pratt & Whitney JT3C-6 turbojets.
Take-off power each (I.S.A., s.l.) : 13,500-lb.
  thrust at 9,950 r.p.m.
Sound suppressors and thrust reversers all engines.
  Water-injection system used.

Weights and loadings

Basic operational : 119,797 lb.
Total fuel : 119,250 lb.
Mfrs. max. payload : 37,315 lb.
Max. take-off : 273,000 lb.
Max. landing : 193,000 lb.
Max. zero fuel : 165,900 lb.
Wing loading (max. t-o. wt.) : 98.6 lb./sq. ft.
Wing loading (max. landing wt.) : 69.6 lb./sq. ft.
Power loading (max. t-o. wt.) : 5.06 lb./lb. thrust.

Performance

Best cost. cruising speed (I.S.A.) : 472 knots at
  30,000 ft. and 220,000 lb.
Approach speed ($1.3 V_{s0}$ at max. landing wt.) :
  128.5 knots.
Take-off field lengths (FAA requirements at max.
  t-o. wt.) :
  At I.S.A. at sea level : 9,380 ft.
Landing field length (FAA requirements at max.
  landing wt.) : 6,410 ft.
Range (take-off to landing), still air, no reserves,
  I.S.A., max. fuel (14,654 imp. gal.) : 4,050 n. mi.
  with 27,840-lb. payload.
Range (take-off to landing), still air, no reserves,
  I.S.A., with max. payload : 3,760 n. mi.

DC-8 Series 30

Powerplants

Four Pratt & Whitney JT4A-11 turbojets.
Take-off power each (I.S.A., s.l.) : 17,500-lb.
  thrust at 9,355 r.p.m. No water injection.

Weights and loadings

Basic operational : 135,378 lb.
Total fuel : 157,000 lb.
Mfrs. max. payload : 34,360 lb.
Max. take-off : 315,000 lb.
Max. landing : 207,000 lb.
Max. zero fuel : 178,200 lb.
Wing loading (max. t-o. wt.) : 109.9 lb./sq. ft.
Wing loading (max. landing wt.) : 72.1 lb./sq. ft.
Power loading (max. t-o. wt.) : 4.5 lb./lb. thrust.

Performance

Best cost. cruising speed (I.S.A.) : 464 knots at
  35,000 ft. and 230,000 lb.
Approach speed ($1.3 V_{s0}$ at max. landing wt.) :
  132.5 knots.
Take-off field lengths (FAA requirements at max.
  t-o. wt.) :
  At I.S.A. at sea level : 9,650 ft.
Landing field length (FAA requirements at max.
  landing wt.) : 6,800 ft.
Range (take-off to landing), still air, no reserves,
  I.S.A., max. fuel (19,480 imp. gal.) : 5,250 n. mi.
  with 25,475-lb. payload.
Range (take-off to landing), still air, no reserves,
  I.S.A., with max. payload : 4,000 n. mi.

DC-8 Series 40

Powerplants

Four Rolls-Royce Conway RCo. 12 turbofans.
Take-off power each (I.S.A., s.l.) : 17,500-lb.
  thrust at 10,060 r.p.m. No water injection.

Weights and loadings

Basic operational : 134,062 lb.
Total fuel : 157,000 lb.

---

5. *A DC-8 Series 20, N6571C, of National
  Airlines.*
6. *Eastern Air Lines' DC-8 Series 20, N8607.*
7. *A DC-8 Series 30 in Northwest markings,
  N804US.*
8. *Pan American's first DC-8 Series 30,
  N800PA.*

5

6

7

8

9

10

11

12

13

Mfrs. max. payload: 36,175 lb.
Max. take-off: 315,000 lb.
Max. landing: 207,000 lb.
Max. zero fuel: 177,100 lb.
Wing loading (max. t-o. wt.): 109.9 lb./sq. ft.
Wing loading (max. landing wt.): 72.1 lb./sq. ft.
Power loading (max. t-o. wt.): 4.5 lb./lb. thrust.

Performance

Best cost. cruising speed (I.S.A.): 465 knots at
   35,000 ft. and 230,000 lb.
Approach speed ($1.3 V_{s_0}$ at max. landing wt.):
   132.5 knots.
Take-off field lengths (FAA requirements at
   max. t-o. wt.):
   At I.S.A. at sea level: 9,650 ft.
Landing field length (FAA requirements at max.
   landing wt.): 6,800 ft.
Range (take-off to landing), still air, no reserves,
   I.S.A., max. fuel (19,480 imp. gal.): 5,905 n. mi.
   with 23,375-lb. payload.
Range (take-off to landing), still air, no reserves,
   I.S.A., with max. payload: 5,310 n. mi.

Engineering summary as for DC-8 Series 50 on p. 62.

---

9. *The Swissair DC-8 Series 30, HB-IDB.*

10. *SAS DC-8 Series 30, OY-KTA.*

11. *The Panair do Brasil DC-8 Series 30, PP-PDT, with a fifth engine pod under the port wing.*

12. *Originally ordered by Pan American, N8274H is one of Panagra's four DC-8 Series 30s.*

13. *A UAT DC-8 Series 30, F-BJLB.*

(continued from p. 57)
engine in intercontinental versions of the DC-8 in 1956. Apart from the engines, the two versions are similar and the modifications described above in respect of the Series 30 are applicable to this version also.

The first order for the Series 40 came from TCA, and two aircraft for this airline were included in the first nine DC-8s built. The first with Conways flew on July 23, 1959, registered N6577C for the duration of the trials, and was followed by N6578C. FAA certification was obtained on March 24, 1960, and TCA began services with this type in April.

Other airlines to order the Series 40 were Alitalia (ten) and Canadian Pacific (four).

Sub-variants of the Series 40 are designated DC-8-41, -42 and -43, equivalent to the respective variants in the DC-8 Series 30. Type approval was obtained on March 23, 1961, April 26, 1960, and February 1, 1961, respectively.

DC-8 Series 50. Final version of the DC-8 passenger airliner, the Series 50 was developed to take advantage of the Pratt & Whitney JT3D turbofan engines. Apart from the power plant, it is basically a Series 30, with all the wing and other modifications developed for that variant.

The turbofan engines, in redesigned nacelles without sound suppressors and with reversers for both gas streams, give a marked improvement in performance and economy. They were first flown in the original DC-8 N8008D on December 20, 1960, which after being used in the certification programme was delivered to National Air Lines in October, 1961.

DOUGLAS DC-8 (U.S.A.)

Most of the certification flying for the Series 50 was done with two of the aircraft destined for KLM, registered N9603Z and N9605Z prior to delivery.

Sub-variants of the Series 50 are designated DC-8-51, -52 and -53. The -51 was approved at a gross weight of 276,000 lb. on October 10, 1961, and is in effect a DC-8-21 with turbofan engines. The DC-8-52 and -53 were both approved on April 28, 1961, at gross weights of 300,000 lb. and 310,000 lb. respectively. Use of 18,000-lb. JT3D-3 turbofans was approved in February, 1962.

Several DC-8 operators specified Series 50s when re-ordering, or changed part of their undelivered orders to this version. United Air Lines had three in a total fleet of forty DC-8s; KLM had seven in a fleet of fourteen; National added six to three Series 20s and Japan Air Lines specified one. Other operators who ordered the Series 50 were Philippine Air Lines (two), Aeronaves (one), Mexicana (one), Iberia (four) and Trans Caribbean (two).

DC-8F Jet Trader. Announced in April, 1961, the DC-8F is a version of the DC-8 Series 50 with provision for freight carrying as well as the customary passenger facilities. After long study of the potential for pure-jet freighters, Douglas concluded that the combination cargo-passenger carrier was a better prospect than more limited all-freight aeroplanes with nose or tail loading.

Compared with the standard DC-8, the Jet Trader has a reinforced floor with built-in track which will accept either seat fittings or rollers, guide rails and holding fixtures for freight pallets. A loading door measuring 85 by 140 in. is provided in the port fuselage side forward.

A removable bulkhead can be located in any one of four positions in the cabin, to separate the freight from the passengers. Thus, the DC-8F is able to carry 183 passengers in an all-economy layout, or 94,668 lb. of freight in an all-cargo layout, or four combinations of freight and passengers between these two limits.

Apart from the fuselage modifications, the Jet Trader is a standard DC-8 Series 50 with JT3D-3 turbofans and gross weight of 315,000 lb. Douglas took the decision to build a DC-8F demonstrator during 1961 and this aircraft was to fly in September, 1962.

Trans-Canada Air Lines ordered four Jet Traders in 1962, for delivery early in 1963. Two of these will be used initially in an all-passenger role and the other two will operate as mixed passenger-cargo carriers on routes across Canada.

---

14. *Japan Air Lines' DC-8 Series 30, JA8001.*
15. *The fourth TCA DC-8 Series 40, CF-TJD.*
16. *Canadian Pacific DC-8 Series 40, CF-CPF.*
17. *A KLM DC-8 Series 30, PH-DCA.*
18. *A KLM DC-8 Series 50, PH-DCM, on lease to VIASA.*

14

15

16

17

18

DOUGLAS DC-8 Series 50

Dimensions
Span: 142 ft. 5 in.
Overall length: 150 ft. 6 in.
Overall height: 42 ft. 4 in.
Gross wing area: 2,771 sq. ft.
Sweepback: 30 deg. on quarter-chord line.
Internal cabin dimensions:
    Length (ex. flight deck): 103 ft.
    Max. width: 140 in.; max. height: 80 in.
Max. usable floor area (ex. flight deck): 1,100 sq. ft.
Max. usable volume (ex. flight deck): 7,940 cu. ft

Accommodation
Typical mixed class: 24F (5 abreast) and 93E (6 abreast).
Max. high density: 177 (6 abreast at 37-in. pitch).
Volume of freight and baggage holds: 1,390 cu. ft.

Powerplants
Four Pratt & Whitney JT3D-3 turbofans.
Take-off power each (I.S.A., s.l.): 18,000-lb. thrust at 10,200 r.p.m.
No water injection or sound suppressors. Thrust reversers on primary and secondary gas streams.

Weights and loadings
Basic operational: 132,325 lb.
Total fuel: 159,300 lb.
Mfrs. max. payload: 34,360 lb.
Max. take-off: 315,000 lb.
Max. landing: 207,000 lb.
Max. zero fuel: 176,500 lb.
Wing loading (max. t-o. wt.): 109.9 lb./sq. ft.
Wing loading (max. landing wt.): 72.1 lb./sq. ft.
Power loading (max. t-o. wt.): 4.37 lb./lb. thrust.

Performance
Best cost. cruising speed (I.S.A.): 473 knots at 35,000 ft. and 220,000 lb.; consumption: 0.042 n. mi./lb.
Approach speed ($1.3 Vs_0$ at max. landing wt.): 132.5 knots.
Take-off field lengths (FAA requirements at max. t-o. wt.):
    At I.S.A. at sea level: 9,300 ft.
Landing field length (FAA requirements at max. landing wt.): 6,550 ft.
Range (take-off to landing), still air, no reserves, I.S.A., max. fuel (19,480 imp. gal.): 6,550 n. mi. with 25,915-lb. payload.

---

19. *Iberia's DC-8 Series 50, EC-ARB.*
20. *A KLM DC-8 Series 50 before delivery, with registration N9605Z.*
21. *Alitalia DC-8 Series 40, I-DIWA.*
22. *The first Trans Caribbean DC-8 Series 50.*

Range (take-off to landing), still air, no reserves, I.S.A., with max. payload: 5,855 n. mi.

DC-8F JET TRADER

Weights and loadings
Basic operational: 130,207 lb.
Total fuel: 159,300 lb.
Mfrs. max. payload: 88,000 lb.
Max. take-off: 315,000 lb.
Max. landing: 240,000 lb.
Max. zero fuel: 224,000 lb.
Wing loading (max. t-o. wt.): 109.9 lb./sq. ft.
Wing loading (max. landing wt.): 86.6 lb./sq. ft.
Power loading (max. t-o. wt.): 4.37 lb./lb. thrust.
Structure. Monocoque elliptical-section fuselage. Close-spaced frames and stringers 75ST, skins 75 and 14ST6. Three-spar wing with roll-tapered 75ST skins.
Fuel system. Four integral tanks each wing, one in centre section. Bag tank in wing fillets. Capacity, 14,654 imp. gal. (Series 10 and 20); 19,481 imp. gal. (Series 30, 40 and 50). Four-point pressure refuelling at 1,200 U.S. g.d.m.
Undercarriage. Four-wheel bogie main units, inwards retracting with castoring rear trucks for taxi-ing. Twin-wheel steerable nosewheel. Track, 20 ft. 10 in. Base, 57 ft. 5 in. Tyre pressure, 148–168 p.s.i.
Flying controls. Interconnected two-piece ailerons each side; outer portions operate through pre-loaded torque link from hydraulically boosted inners at low speeds only. Hydraulically boosted rudder. Manual elevator with aerodynamic tabs, and variable-incidence tailplane. Sperry S.P.30 autopilot.
Flaps, air brakes. Double-slotted flaps in four portions. Two fixed leading-edge slots each side, inboard of pylons. Four upper wing spoilers linked with nosewheel to destroy lift.
Cabin conditioning. Four turbocompressors powered by engine-bleed air for identical duplicated system. Freon vapour-cycle refrigeration. Differential, 8.77 p.s.i.
Hydraulics. Main 3,000-p.s.i. system from two engine-driven pumps and electrically driven pumps for auxiliary system, for u/c, flaps, power controls, nosewheel steering and brakes; separate electrically powered system for spoilers.
Electrics. Four 20-kVA. engine-driven generators for three-phase 115/200-V., 400-cycle constant frequency a.c. and four 50-amp. transformer-rectifiers for 28-V. d.c.
De-icing. Wings, tail unit and intakes by engine-bleed air; Nesa glass screen and air-blast rain removal.

FOKKER **FRIENDSHIP** (Netherlands)

A Bonanza Air Lines' F-27A, N145L, from the Fairchild production line.

BEFORE THE Second World War, the name of Fokker was among the most respected in commercial aviation. A long series of Fokker airliners had been produced for use not only by KLM, the Dutch national airline, but also for many of the world's leading airlines. The three-engined Fokker F.VII/3m in particular established a lasting reputation as one of the best airliners of its time.

Fokker's factory in Amsterdam was destroyed during the war and its staff dispersed, but a rapid recovery was made in 1945 and a new design team began work. Project work on new airliners was given high priority.

During 1950, studies of a new type began on the basis of a Dakota replacement. The Fokker factories had been responsible since 1945 for the overhaul and conversion of Dakotas, and this work brought the company into close contact with many European and other airlines. An extensive traffic study was made to discover the size of aeroplane which might be needed, and DC-3 operators were asked to give an outline of their likely requirements for a replacement aircraft. Fokker also purchased from the Boeing company in America the project designs of an aircraft known as the Boeing 417, together with the results of a market survey made by Boeing of the DC-3 replacement prospects. The Boeing 417 had a circular-section cabin for twenty-four passengers and two Wright Cyclones on a high wing.

One of the earliest projects in the series leading to the F-27 was known as the Fokker Type 275, dated August, 1950. This had a shoulder-mounted wing and a circular-section fuselage accommodating thirty-two passengers. Other features were a single tail unit, tricycle undercarriage and two Rolls-Royce Dart R.Da.3 engines.

1. *Fokker's second F-27, PH-NVF.*
2. *The second Fairchild F-27, N2027, at take-off.*
3. *First F-27 for the first U.S. operator, West Coast Airlines.*
4. *CF-QBR, Fairchild-built F-27 for Quebecair.*
5. *An Avensa F-27, YV-C-EVH, built by Fairchild.*

In most respects, the original Type 275 resembled the F-27 as eventually built, but many other possible layouts were also considered. These included single-engined types, piston, turbojet and turboprop power plant and high- and low-wing arrangements.

By 1952, most of the basic design decisions had been made. The aeroplane which Fokker wished to build would seat from thirty-two to forty passengers, and would carry its maximum payload about 300 miles. Compared with the DC-3, it would need to be pressurized, have a higher cruising speed and a nosewheel undercarriage, but be no more expensive to operate. To meet these requirements, the aeroplane would have a high wing of high aspect ratio, and two Dart engines in underslung nacelles.

An extensive programme of wing tunnel testing was conducted, principally at the National Aeronautical Laboratory (NVL), and in 1953 the Netherlands Aircraft Development Board undertook to finance the construction of three (later four) prototypes. Of the four, two were flying prototypes and the other two airframes were used for static and fatigue testing on the ground.

During 1952, Fairchild Airplane Division of Hagerstown, Maryland, had obtained a licence from Fokker to produce the S.11, S.12, S.13 and S.14 series of trainers and although none of these was built in America the agreement led to Fairchild acquiring an option on a further licence to produce the F-27 in the U.S.A. This option was taken up after the first flight of the F-27 at Schiphol and a new licence agreement was signed between the two companies on April 26, 1956.

Fokker flew their first F-27 (PH-NIV) on November 24, 1955, and the second (PH-NVF) followed on January 31, 1957. Early flight trials showed that the double-slotted flaps fitted on the first aircraft were not needed to achieve the design performance, and these were replaced by single-slotted type as the production standard. The first prototype also differed from all others, including the second, in the fuselage length, which was three feet less to start with. The longer fuselage allowed an increase in the basic seating from thirty-two to thirty-six. Dart 507s were fitted for the first flights of PH-NIV, with the production-type Dart 511s substituted later.

FOKKER FRIENDSHIP (Netherlands)

In the U.S., Fairchild converted the Fokker drawings to American engineering standards and incorporated certain new features in the standard model. These included a Stratos-designed air-conditioning system, integral stairs in the main passenger door and provision for weather radar in a lengthened nose. (eventually adopted also by Fokker as standard). The basic seating was for forty passengers, and Fairchild increased the standard fuel capacity from 952 to 1,320 U.S. gallons with another 340 U.S. gallons in optional centre section bag tanks.

The first Fairchild F-27 (N1027) was flown at Hagerstown on April 12, 1958, followed by the second (N2027) on May 23. Both were eventually sold to airlines after they had completed their development programmes.

Fokker adopted the name Friendship for the F-27, and uses Series numbers (100, 200, 300, 400 and 500) to identify the principal variants. Fairchild uses the F-27 designation without the name and identifies the variants by suffix letters. The subsequent history of the design is related under the variant headings which follow.

F-27 Series 100 (Fairchild F-27). This is the basic version built both by Fokker and by Fairchild. With Dart R.Da.6 Mk. 511 engines, it went into production with a gross weight of 35,700 lb. As already noted, the second Fokker prototype was to production standard so far as general dimensions and equipment were concerned. The first two Fairchild F-27s were genuine production models, and were used to obtain certification (on July 16, 1958) which cleared the way for deliveries to begin. Orders had been obtained by Fairchild from several of the U.S. local airlines and one of these, West Coast Airlines, became the first to operate the F-27 on September 28, 1958. This was the first time a turbine-engined transport of American manufacture had been operated in airline service.

Fokker did not fly their first production F-27 until March 23, 1958. This was completed in Aer Lingus colours, and the Irish airline became the first operator of the Dutch-built transport in December, 1958. Type approval was obtained from the Dutch and U.S. authorities on October 29, 1957.

At the thirty-first production line position,

Fokker introduced the larger fuel tanks which had been standard on the Fairchild F-27 from the beginning. External tanks were also developed to permit long-range ferry flights to be made; with a capacity of 200 imp. gallons each they gave a still air range of up to 1,850 naut. miles.

In September, 1959, the gross weight of the Series 100 aeroplane increased to 37,500 lb. Two years later, in mid-1961, the weight was again increased, to 39,000 lb., and for those aircraft fitted with Dart 511-7E engines (with Nimonic 105 turbine blades) this weight could be used in air temperatures up to 31° C. at sea level.

The full list of airline orders for the Fokker-built F-27 Series 100, with dates of first service, is as follows: Aer Lingus, seven (December, 1958); Braathens SAFE, four (January, 1959); East West Airlines, one (August, 1959); Philippine Airlines, five (1960); THY, five (1960); NZNAC, eight (December, 1960); Indian Airlines, ten (1961); ALA, three (1962); Fujita, two (1962).

Fairchild airline orders are: West Coast, six (September, 1958); Quebecair, three (October, 1958); Avensa, five (October, 1958); Piedmont, eight (November, 1958); Trans Mar de Cortes, one (1959); Ozark, three (1960); Aloha, six (June, 1959); THY, five (1960).

F-27 Series 200 (F-27A). To provide better take-off performance in high temperatures or at high-altitude airports, Fokker developed this second variant of the Friendship with the more powerful Dart R.Da.7 Mk. 528 engines. Other than this change, which gave higher cruising speeds as well as better take-off, for the loss of some range, the Series 200 was identical with the Series 100.

Fokker substituted Dart 528s for 511s in the prototype PH-NIV, and Fairchild similarly

6. *Piedmont F-27, N2701R, by Fairchild.*
7. *The first F-27B with large freight door, N4903, of Northern Consolidated Airlines.*
8. *Fokker-built for Aer Lingus, the F-27 EI-AKE.*
9. *Braathen's Fokker F-27 in pre-delivery markings as PH-FAD.*
10. *A Fairchild-built F-27A, N2770R, for Pacific Air Lines.*

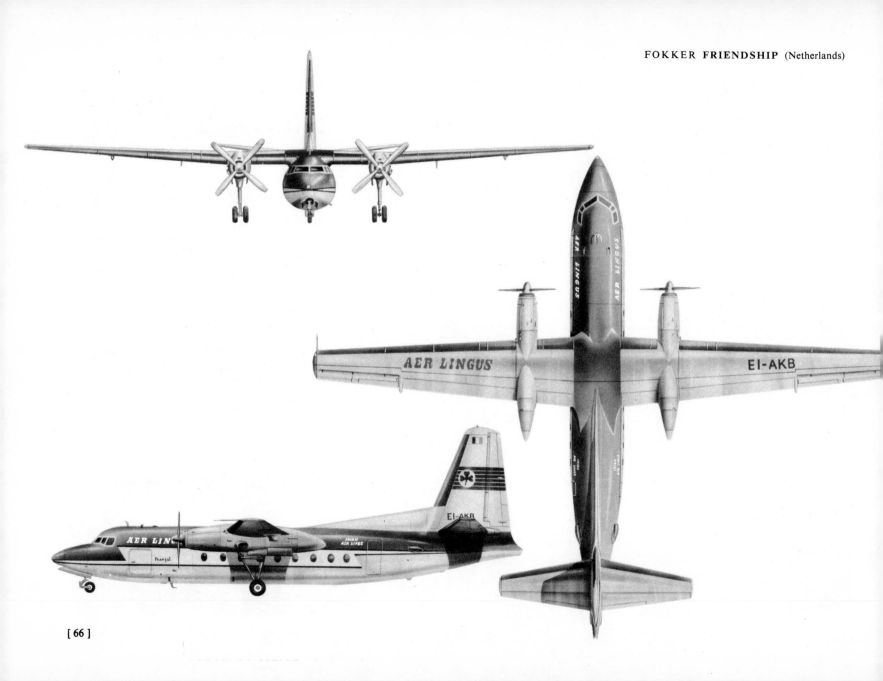

FOKKER FRIENDSHIP (Netherlands)

FOKKER FRIENDSHIP (Netherlands)

re-engined their first aircraft which became an F-27A when it first flew with Dart 528s. The latter aircraft was eventually delivered to the Ecuadorian airline AREA after type approval for the F-27A was obtained on December 31, 1958.

Originally certificated at a weight of 35,700 lb., the Series 200 was cleared for operations at 37,500 lb. in September, 1959, and in mid-1961 a further increase was made to 42,000 lb. For aircraft with Dart 528-7Es (with Nimonic 105 turbine blades) the higher weight could be used in temperatures up to 40° C.

Fokker have sold Series 200s to the following airlines: Trans-Australia Airlines, ten (May, 1959); Ansett-ANA (including Airlines of N.S.W. and Queensland Airlines), nine (November, 1959); Mac.Robertson Miller Airlines, one (December, 1959); Pakistan International, four (February, 1961); All Nippon Airways, ten (May, 1961); Sudan Airways, three (1962); DETA, three (1962); DTA, two (1962); East African, three (1963); Nigeria Airways, five (1963); Malayan Airways, five (1963).

Fairchild delivered F-27As as follows: Bonanza Airlines, ten (March, 1959); Pacific Airlines, six (April, 1959); Wien Alaska, two (July, 1959); Aerovias Ecuatorianas, one (1960).

F-27 Series 300 (F-27B). Sometimes called the Combiplane, the Series 300 is the same as the Series 100 but has a large freight-loading door in the forward fuselage, and a suitable cabin floor for freight carrying. A removable bulkhead makes this version suitable for mixed passenger-freight operations.

The first examples of this version came from the Fairchild line and were delivered to Northern Consolidated under the designation F-27B in November, 1958. This Alaskan operator received three following certification on October 25, 1958.

Fokker built two Series 300s for KLM—which was the first operator of any to order the Friendship, but which has chartered its two aircraft to an Iranian oil company. The German charter operator LTU took delivery of a Series 300 early in 1962 after using the

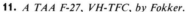

11. *A TAA F-27, VH-TFC, by Fokker.*
12. *Aloha Airlines' Fairchild F-27, N5094A.*
13. *The Wien Alaska F-27, N2708R, by Fairchild.*
14. *A TAA F-27A leased to East West and re-registered VH-EWA.*
15. *An Ansett-ANA F-27A in the markings of the subsidiary company, Airlines of NSW.*
16. *Ozark Air Lines' Fairchild-built F-27, N4300F.*

11

12

13

14

15

16

[67]

17

18

19

20

second Fokker prototype on charter between February and October, 1961. Another was ordered at the end of 1961 by Luxair, and this Luxembourg operator also arranged to lease the second prototype for a short time in 1962.

F-27 Series 400. This variant is identical with the Series 300 apart from having the Dart 528 engines. It can therefore be regarded as the Combiplane version of the Series 200. One example has been delivered by Fokker to Pakistan International Airlines.

17. *In Ansett-ANA markings, this Fokker F-27A carries the pre-delivery marking PH-FBA.*
18. *The Mac.Robertson Miller F-27A, RMA Swan.*
19. *New Zealand National Airways Corpora-*
tion's Fokker F-27, ZK-BXA, Kuaka.
20. *The Fairchild F-27F demonstrator N27C showing the integral airstairs.*
21. *Indian Airlines' Fokker F-27, VT-DMD.*
22. *Pakistan International Airlines' Fokker F-27, AP-ALN.*

21

22

FOKKER FRIENDSHIP (Netherlands)

F-27 Series 500. During 1961, Fokker announced a "long-fuselage" version of the Friendship. This was developed to allow airline operators to take advantage of the approved higher operating weights for the Friendship in terms of greater payload. The principal change is a 4-ft. 11-in. "stretch" of the fuselage, both ahead and behind the wing. The usable volume of the fuselage is thereby increased by 230 cu. ft. to 2,360 cu. ft.

The Series 500, which had not been ordered by any airline up to the spring of 1962, has the large forward freight loading door as standard. Some modifications are also made to the undercarriage, which is a few inches longer and is moved two inches aft.

Gross weight of the Series 500 is 42,000 lb. and with Dart 528s it has an estimated cruising speed of 252.5 knots at 38,000 lb. at 20,000 ft. Accommodation would be for up to fifty-two passengers in a high-density arrangement.

F-27F. In conjunction with Fokker, Fairchild developed this refinement of the F-27A during 1961, and put the type back into production after a lapse of a year. One external change is deletion of the cabin conditioning air intakes on the rear fuselage, which are replaced by a flush intake in the base of the fin.

On February 24, 1961, FAA certification of the new model was obtained, at a gross weight of 38,500 lb. with Dart 529-7E engines. These engines are rated at 2,190 e.h.p., a little more powerful than the Dart 528-7E. The gross weight was to go up later in 1962 to 42,000 lb. when a fuel injection system had been introduced to meet FAA requirements.

F-27G. Also primarily a Fairchild variant, this is a mixed passenger-freight version of the F-27F, with large forward door and provision for extra fuel. The maximum fuel capacity is 2,090 U.S. gallons, compared with the standard 1,320 U.S. gallons and maximum 1,680 U.S. gallons in earlier versions.

F-27H. This designation is reserved by Fairchild for a short-field version of the F-27. With double-slotted flaps and a variable-incidence tailplane, it is similar to a Fokker development model which has been referred to as the F-27S. Tests were made late in 1961 with the first Fokker prototype, with military as well as civil applications in view.

F-27M Troopship. Fokker obtained an order for twelve F-27s from the Royal Netherlands Air Force, and has delivered nine of these in Troopship configuration with provision for paratroop dropping and the carriage of military supplies. Preliminary paradrops were made from the second prototype before the military F-27 was ordered.

23

24

25

26

27

FOKKER F-27

Dimensions

Span: 95 ft. 2 in.
Overall length: 75 ft. 9 in.
Overall height: 27 ft. 11 in.
Gross wing area: 754 sq. ft.
Sweepback: Nil.
Internal cabin dimensions:
　Length (ex. flight deck): 47 ft. 6 in.
　Max. width: 100.5 in.; max. height: 79 in.
Max. usable floor area (ex. flight deck): 312 sq. ft.
Max. usable volume (ex. flight deck): 2,130 cu. ft.

Accommodation

Typical standard: 40–48 (4 abreast at 35¼–29¼-in. pitch) or: 44–48 (4 abreast at 35¼–33-in. pitch).

Max. high density: 52 (4 abreast at 30-in. pitch).
Volume of freight and baggage holds: 185 cu. ft.

F-27 Series 100

Powerplants

Two Rolls-Royce Dart 511-7E turboprops.
Take-off power each (I.S.A., s.l.): 1,710 e.h.p. at 14,500 r.p.m.

23. *An F-27A for Philippine Air Lines, PI-C501.*
24. *One of the ten THY F-27s, TC-TEK.*
25. *For All Nippon Airways, the F-27, JA8601.*
26. *Luft-transport Union's F-27, D-BAKU.*
27. *The first F-27A for Sudan Airways.*

FOKKER FRIENDSHIP (Netherlands)

Water injection system installed.
Rotol four-blade 12-ft. reversing propellers.

Weights and loadings

Basic operational: 23,000 lb.
Total fuel: 9,000 lb.
Mfrs. max. payload: 10,194 lb.
Max. take-off: 39,000 lb.
Max. landing: 37,500 lb.
Max. zero fuel: 35,200 lb.
Wing loading (max. t-o. wt.): 51.7 lb./sq. ft.
Wing loading (max. landing wt.): 50.0 lb./sq. ft.
Power loading (max. t-o. wt.): 12.2 lb./e.h.p.

Performance

Best cost. cruising speed (I.S.A.): 235.5 knots at
20,000 ft. and 35,000 lb.; consumption: 156.5
imp gal./hr.
Approach speed ($1.3 Vs_0$ at max. landing wt.):
88 knots.
Take-off field lengths (ICAO-PAMC require-
ments at max. t-o. wt.):
At I.S.A. at sea level: 5,720 ft.
At I.S.A. +15° C. at sea level: 6,200 ft.
Landing field length (ICAO-PAMC requirements
at max. landing wt.): 3,160 ft.
Range (take-off to landing), still air, no reserves,
I.S.A., max. fuel (1,150 imp. gal.): 1,120 n. mi.
with 6,850-lb. payload at 235 knots (mean) at
20,000 ft. (mean).
Range (take-off to landing), still air, no reserves,
I.S.A., with max. payload: 550 n. mi. at 235
knots (mean) at 20,000 ft.

F-27 Series 200

Powerplants

Two Rolls-Royce Dart 528-7E turboprops.
Take-off power each (I.S.A., s.l.): 2,105 e.h.p. at
14,500 r.p.m.

Weights and loadings

Basic operational: 23,526 lb.
Total fuel: 9,000 lb.
Mfrs. max. payload: 10,197 lb.
Max. take-off: 42,000 lb.
Max. landing: 37,500 lb.
Max. zero fuel: 35,700 lb.
Wing loading (max. t-o. wt.): 55.7 lb./sq. ft.
Wing loading (max. landing wt.): 50.0 lb./sq. ft.
Power loading (max. t-o. wt.): 11.0 lb./e.h.p.

Performance

Best cost. cruising speed (I.S.A.): 262 knots at
20,000 ft. and 38,000 lb.; consumption: 181
imp. gal./hr.
Take-off field lengths (ICAO-PAMC require-
ments at max. t-o. wt.):
At I.S.A. at sea level: 5,600 ft.
At I.S.A. +15° C. at sea level: 6,100 ft.
Landing field length (ICAO-PAMC requirements
at max. landing wt.): 3,160 ft.
Range (take-off to landing), still air, no reserves,
I.S.A., max. fuel (1,150 imp. gal.): 1,100 n. mi.
with 9,150-lb. payload at 262 knots (mean) at
20,000 ft. (mean).
Range (take-off to landing), still air, no reserves,
I.S.A., with max. payload: 960 n. mi. at 262
knots (mean) at 20,000 ft.

Structure. Semi-monocoque oval-section fuselage.
Two-spar wing comprising centre section,
outer panels and tips, with Redux bonding.
Principal materials 75ST6 and 24ST4.

Fuel system. Two integral tanks in wings. Capacity, 795 imp. gal. Gravity refuelling (pressure
optional) at 80 g.p.m.

Undercarriage. Twin main wheels and steerable
nose oleos. Track, 23 ft. 7½ in. Base, 28 ft. 8 in.
Tyre pressure, 80 p.s.i.

Flying controls. Manual, aerodynamically
balanced ailerons with trim, spring and servo
tabs; manual, aerodynamically balanced elevator and rudder, the latter with geared and
trim tabs. Smiths S.E.P. 2 autopilot.

Flaps, air brakes. Single-slotted flaps in one piece
each side, electrically actuated. No air brakes.

Cabin conditioning. Two engine-driven compressors. Differential, 4.16 p.s.i.

Pneumatics. Two independent 3,300-p.s.i. pneumatic systems with engine-driven compressors
for u/c, brakes and nosewheel steering.

Electrics. 24–28-V. d.c., from two engine-mounted
generators and secondary three-phase 115-V.
400-cycle a.c. from two invertors.

De-icing. Wing and tail unit by rubber boots.
Electrical de-icing of intakes, propellers,
spinners and windshield.

HANDLEY PAGE HERALD

THE HERALD feeder-liner is one of the very
few examples of an aeroplane designed for
piston engines which has been successfully
converted for turboprop power. A parallel is
to be found in the Nord Super Broussard
(p. 84).

The decision to build a short-haul airliner
was taken by Handley Page in the early 'fifties,
and an extensive market survey was conducted to establish the basic design features.
This survey, in the course of which the company obtained the opinions of nearly every
airline in Europe, Asia, Africa, Australasia
and South America, led to the development of
a pressurized aeroplane with maximum accommodation for forty-four passengers. A high-
wing position was adopted for ease of loading,
good passenger view from the cabin and
other reasons, and four Alvis Leonides Major
piston radial engines were chosen to comprise
the power plant.

The project design, detail design and production of the new airliner were assigned to
the Reading works of Handley Page and the
designation HPR.3 was allocated to the
design. By the end of 1954, all detail design
was virtually complete and construction of
two prototypes was well advanced. From the
start, the project was undertaken as a private
venture wholly financed by Handley Page.

During 1954, full details of the Herald were
given for the first time. The aircraft had a
gross weight of 34,000 lb. and a payload of
10,700 lb., which could be carried for 348 miles.
With full tanks, a range of 1,420 miles could
be flown with a 6,930-lb. payload. The span
was 95 ft. 0 in. and the length 70 ft. 3 in.
Towards the end of the year, Queensland
Airlines announced its intention to order
Heralds and this was followed in February,
1955, by an order from Australian National
Airways. The total ordered by the two Australian airlines was twenty-four. Five were
ordered by Lloyd Aereo Colombiano in
March, 1955, and a provisional order was
also placed by Air Kruise, a British independent airline.

The first Herald, G-AODE, flew from
Woodley Aerodrome on August 25, 1955, in
Queensland markings, and was followed by
the second prototype, G-AODF, in August,
1956. While these two aircraft proceeded with

HANDLEY PAGE **HERALD** (Great Britain)

The first production Handley Page Herald, G-APWA, in BEA markings.

certification trials, a production line was laid down and material was purchased for a batch of twenty-five. Before the first of these reached final assembly, however, a change in the attitude of potential customers and other operators became apparent to the designers. The success of the Viscount and development of economical small gas-turbine engines made the piston-engined Herald much less attractive than it had been, and the two Australian operators in particular refused to accept the Herald in this form.

In May, 1957, therefore, Handley Page announced plans to produce an alternative version with two Rolls-Royce Darts. The Leonides version was subsequently dropped completely.

The Dart Herald, with the design designation HPR.7, needed comparatively little modification. Principal changes were a new centre section carrying the nacelles; a modified outer-wing structure with integral fuel tanks; and a 20-in. lengthening of the fuselage. The engine nacelles were substantially the same as those designed by Vickers for the Viscount, and the R.Da.7 version of the Dart was chosen.

The first Herald, G-AODE, was withdrawn from flight testing in June, 1957, in order to be converted to the new standard. It re-emerged at Woodley in time to resume its flight tests, as the prototype Dart Herald, on March 11, 1958. Preliminary experience was promising but on August 30, 1958, while *en route* to the SBAC Display at Farnborough, the aircraft suffered an engine failure followed by fire and was destroyed in the forced landing which followed.

Consequently, the second Herald, G-AODF, was modified to have Darts, flying for the first time on December 17, 1958. This aircraft was used for the bulk of the certification flying, much of which had been completed by the first Herald before it crashed. A preliminary C. of A. had been obtained, in fact, on April 22, 1958.

On June 11, 1959, the British Government announced it would buy three Heralds for use by BEA on the Scottish Highlands and Islands routes. This was the first firm order for the Herald, and the prototype G-AODF was repainted in BEA colours.

Herald Series 100. Production models to the original specifications now carry this designation. The first to fly was G-APWA, on October 30, 1959, also in BEA colours. This aircraft is a Handley Page demonstrator, the next three on the line, G-APWB, -WC and -WD, being those for BEA. During 1961, after Jersey Airlines had ordered six of the slightly larger Series 200 Herald, the first two Series 100s, G-APWA and G-APWB, were leased to Jersey Airlines. BEA accepted delivery of G-APWC in January, 1962, with G-APWB and G-APWD to follow.

Herald Series 200. In September, 1960, Jersey Airlines ordered six Heralds of a new series with longer fuselage and higher gross weights. Designated Series 200, this version retains most of the characteristics of the Herald 100. The prototype G-AODE was converted to the new standard and resumed its flight trials on April 8, 1961. In August, it was re-registered G-ARTC, fitted with a nose radome and painted in the markings of Maritime Central Airways, a Canadian operator which had ordered two Heralds earlier in the year. The C. of A. for this variant was obtained on June 1, 1961.

The first production Herald 200 was G-APWE, first flown in December 13, 1961, and delivered to Jersey Airlines in January, 1962. In February, the first Series 200 for the Canadian operator Nordair was delivered,

1. *The first Herald, G-AODE, with four Leonides Major engines.*
2. *The second Herald, G-AODF, after conversion to Dart power.*
3. *In Jersey Airlines' markings, the first production Herald G-APWA.*
4. *After conversion to a Series 200, the second Herald G-AODF.*
5. *The first production model Herald 200, G-APWE, for Jersey Airlines.*

followed by the second in March. An order for two Heralds from Itavia was announced in September, 1961, and in 1962 Caledonian Airways and Eastern Provincial Airlines, in Canada, ordered one and two respectively. One

Herald 200 has been purchased by British United Airways in executive configuration.

6. *The first Herald for Nordair, CF-NAC, at take-off.*

6

HANDLEY-PAGE **HERALD 200**

Dimensions

Span: 94 ft. 9 in.
Overall length: 75 ft. 5 in. (Series 100, 71 ft. 11 in.)
Overall height: 24 ft. 0 in.
Gross wing area: 886 sq. ft.
Sweepback: Nil.
Internal cabin dimensions:
    Length (ex. flight deck): 34 ft. 7 in.
    Max. width: 104 in.; max. height, 77 in.
Max. usable floor area (ex. flight deck): 290 sq. ft.
Max. usable volume (ex. flight deck): 2,415 cu. ft.

Accommodation

Typical coach class: 50 (4 abreast at 30.5-in. pitch).
Max. high density: 56 (4 abreast at 30.0-in. pitch).
Volume of freight and baggage holds: 285 cu. ft.

Powerplants

Two Rolls-Royce Dart 527 (R.Da.7/1) turbo-props.
Take-off power each (I.S.A., s.l.): 2,105 e.h.p. at 15,000 r.p.m.
Water-injection system fitted. Rotol 4-blade 12-ft. 6-in. reversing propellers.

Weights and loadings

Basic operational: 24,958 lb.
Total fuel: 8,640 lb.
Mfrs. max. payload: 11,746 lb.
Max. take-off: 43,000 lb.
Max. landing: 39,500 lb.
Max. zero fuel: 37,000 lb.
Wing loading (max. t-o. wt.): 48.5 lb./sq. ft.
Wing loading (max. landing wt.): 44.6 lb./sq. ft.
Power loading (max. t-o. wt.): 10.2 lb./e.h.p.

Performance

Best cost. cruising speed (I.S.A.): 238 knots at 15,000 ft. and 36,000 lb.; consumption: 0.136 n. mi./lb.
High-speed cruise: 237 knots at 15,000 ft.
Long-range cruise: 204 knots at 23,000 ft.
Approach speed (1.3V_{S_0} at max. landing wt.): 92 knots.

Take-off field lengths (BCAR requirements at max. t.o. wt.):
    At I.S.A. at sea level: 5,450 ft.
    At I.S.A. +15° C. at sea level: 5,875 ft.
    At I.S.A. at 5,000 ft.: 6,500 ft.
Landing field length (BCAR requirements at max. landing wt.): 3,750 ft.
Range (take-off to landing), still air, no reserves, I.S.A., max. fuel (1,080 imp. gal.): 1,420 n. mi. with 9,000-lb. payload at 230 knots (mean) at 23,000 ft. (mean).
Range (take-off to landing), still air, no reserves, I.S.A., with max. payload: 950 n. mi.

Structure. Stressed-skin compound-radius section fuselage. Three-spar centre section and two-spar outer wings with detachable tips. Spot-welding and sandwich construction throughout airframe.

Fuel system. Four bag tanks in centre section and integral tank in each outer wing. Capacity, 1,080 imp. gal. Gravity refuelling (pressure optional) at 120 g.p.m.

Undercarriage. Twin-wheel nose and main units. Steerable nosewheel. Track, 22 ft. Base, 25 ft. 10 in. Tyre pressure, 58 p.s.i.

Flying controls. Rod-, cable- and chain-operated. Spring-tabbed ailerons and rudder, spring and geared tabs on elevators.

Flaps, air brakes. Fowler-type flaps in one piece each side. No air brakes.

Cabin conditioning. Two engine-driven compressors. Differential, 3.35 p.s.i.

Hydraulics. 3,000-p.s.i. system supplied by two engine-driven pumps, for u/c, flaps, nosewheel steering, brakes.

Electrics. Two 28-V. 6-kW. d.c. generators (one each engine). Four 24-V. batteries for starting, general and emergency use.

De-icing. Wing, fin and tailplane by hot air from engine exhaust heat exchanger. Liquid system for windscreen.

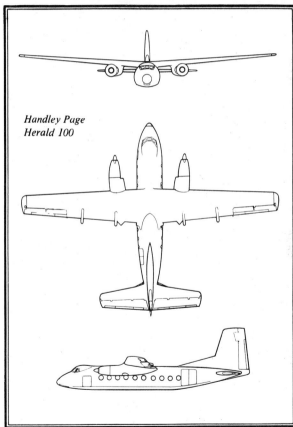

Handley Page Herald 100

ILYUSHIN **Il-18** (U.S.S.R.)

An Aeroflot Il-18, CCCP-75686, landing at Le Bourget, Paris.

ILYUSHIN Il-18 (U.S.S.R.)

COMPARABLE IN SIZE and performance to the Vickers Vanguard and Lockheed Electra, the Ilyushin Il-18 has proved to be the most "exportable" of the range of Russian turbine-engined airliners. By the spring of 1962 it was in service with the airlines of at least nine countries outside Russia and although it suffered from the use of engines with no commercial operating background, it was generally regarded as being as good as its Western contemporaries.

The Il-18 was first flown in mid-1957, the pilot being the veteran Soviet flyer Vladimir Kokkinaki. This prototype, registered CCCP-L5811 and carrying the type-name Moscow on the fuselage, was shown publicly, together with the Antonov An-10 (p. 2) and Tupolev Tu-110 at Vnukovo Airport in Moscow in July the same year. It was reported that only three test flights had been made at this time.

The first aircraft was powered by 4,000-e.h.p. Kuznetsov NK-4 turboprops, and these engines were used in half the preliminary batch of twenty Il-18s, which were built at a factory at Vnukovo. The other ten aircraft had Ivchenko AI-20 turboprops of similar power, and the latter engines were adopted as the production standard after field proving trials of the two types. Apart from this change,

the Il-18 went into production virtually unchanged from the prototype.

Aircraft from the preliminary production batch were used, from the early months of 1958 onwards, for extensive proving flights of various kinds. One of the first of these to receive publicity was a journey from Moscow by way of Irkutsk, to the "North Pole 6" drifting station, and back to Moscow by way of Tiksi. The flying time for this 10,662-mile journey was 27 hr. 30 min. at an average speed of 388 m.p.h., and the flight was completed on March 22, 1958.

Later the same year, Vladimir Kokkinaki was in command of an Il-18 for the first of a series of flights which have broken the official FAI class records for speed and altitude with various payloads. On November 14, an altitude of 40,915 ft. was achieved carrying a

---

1. *Polish Air Lines LOT has three Il-18s, including SP-LSB illustrated.*
2. *The first Malev Il-18, HA-MOA.*
3. *One of two Air Guinée Il-18s, 3X-NZE.*
4. *Ghana Airways received eight Il-18s; 9G-AAJ is shown.*
5. *A Tarom Il-18, YR-IMA.*
6. *The first CSA Ilyushin Il-18 OK-NAA.*

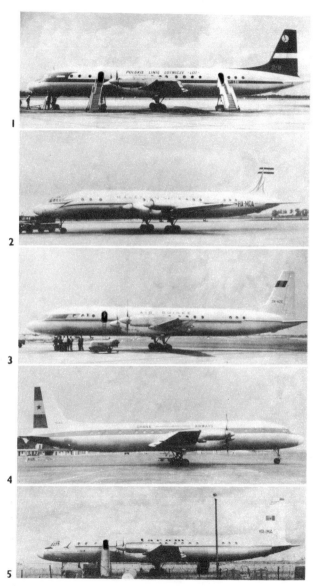

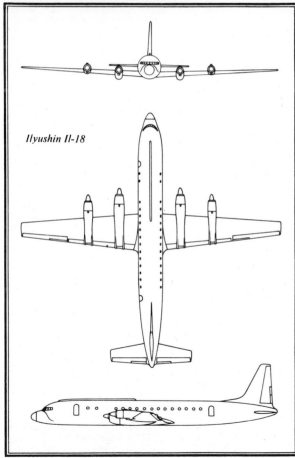

Ilyushin Il-18

payload of 15,000 kg. (33,070 lb.) and the following day, 10,000 kg. (22,046 lb.) was carried to 43,156 ft. The 5,000-kg. (11,023-lb.) record was set at 43,550 ft. on November 11.

These records were established while the Il-18 was still engaged in route proving in the Ukraine, carrying freight and mail. On April 20, 1959, it became the first Russian turbo-prop in scheduled passenger service, flying at first between Moscow and Adler, and Moscow and Alma Ata. Since that date, it has been used in increasingly large numbers by Aeroflot as the principal equipment on the trunk and major feeder routes.

Various cabin layouts are available in the Il-18, those most commonly used by Aeroflot providing for seventy-three to eighty-four passengers in two cabins. The forward cabin seats fifteen to twenty in five-abreast seat rows, and is separated from the main cabin by a large galley and baggage compartment. Maximum layouts, using a rather cramped six-abreast layout, provide for as many as 120 passengers. A de luxe or executive layout has also been illustrated in Russian brochures, but is not known to be in service.

With the Il-18 established in service, Kokkinaki resumed his record flights in August, 1959. On the 19th, he made a flight over a 2,000-km. (1,243-mile) circuit carrying a 15,000-kg. (33,070-lb.) payload at an average speed of 447.07 m.p.h. This represented records also in the classes for 1,000-kg., 2,000-kg., 5,000-kg. and 10,000-kg. payloads.

On November 15, 1959, a payload of 20,000 kg. (44,092 lb.) was lifted to a height of 39,370 ft., and on February 2, 1960, a flight round a 5,000-km. circuit (3,107 miles) was

7. *Deutsche Lufthansa's Ilyushin Il-18 DM-STB.*

made at an average speed of 430.5 m.p.h. This latter flight represented records for nil, 1,000-kg., 2,000-kg., 5,000-kg. and 10,000-kg. payloads.

More noteworthy flights followed as the Il-18 began to visit other parts of the world, carrying Russian leaders and delegations. Examples of these flights were the 13-hr. 56-min., two-sector flight from Moscow to Washington by way of Keflavik; and the 21-hr. 5-min., three-sector flight from Moscow to Mexico by way of Keflavik and Halifax.

The Il-18 began to appear in the markings of airlines other than Aeroflot in 1960, with the delivery of small fleets to the East European Communist-block countries. The Czech airline CSA received its first Il-18 in January, 1960, and began scheduled operations in April. By the end of 1960, the type was also in service with the East German DLH, Malev in Hungary, Tarom in Rumania and the Civil Administration Air Fleet in China. In 1961, deliveries were made to LOT in Poland and to Air Guinée, Air Mali and Ghana Airways. Tabso, in Bulgaria, received its first in 1962.

Complete details of the fleets of these foreign operators are not known. The available facts are as follow: CSA, six (OK-NAA, OK-NAB, OK-OAC, OK-OAD, OK-PAE, OK-PAF); DLH, three (DM-STA, DM-STB, DM-STC); Malev, four (HA-MOA, HA-MOB, HA-MOC, HA-MOD); Tarom, four (YR-IMA, YR-IMB and two others); LOT, three (SP-LSA, SP-LSB, SP-LSC); Ghana Airways, eight (9G-AAI to 9G-AAN, 9G-AAX and 9G-AAY); Air Guinée, two (3X-NZE and 3X-CKN) and Air Mali, two (TZ-ABE and one other).

In the spring of 1962, Aeroflot introduced a new version known as the Il-18E, which has integral fuel tanks in the wings and considerably greater range.

ILYUSHIN Il-18

Dimensions

Span: 122 ft. 8½ in.
Overall length: 117 ft. 9 in.
Overall height: 33 ft. 4 in.
Gross wing area: 1,507 sq. ft.
Sweepback: Nil.
Internal cabin dimensions:
  Length (ex. flight deck): about 85 ft. 0 in.
  Max. width: 127 in.; max. height: 73 in.
Max. usable volume (ex. flight deck): 8,475 cu. ft.

Accommodation

Typical tourist class: 75 (5 abreast).
Max. high density: 120 (6 abreast).
Volume of freight and baggage holds: 953 cu. ft.

Powerplants

Four Ivchenko AI-20 turboprops.
Take-off power each (I.S.A., s.l.): 4,000 e.h.p.
Four-blade AB-68I 14-ft. 9-in. reversing propellers.

Weights and loadings

Basic operational: 61,730 lb.
Total fuel: 42,770 lb.
Mfrs. max. payload: 30,865 lb.
Max. take-off: 135,600 lb.
Wing loading (max. t-o. wt.): 89.1 lb./sq. ft.
Power loading (max. t-o. wt.): 8.5 lb./e.h.p.

Performance

Typical cruising speed (I.S.A.): 330 knots at 26,250 ft.
High-speed cruise: 350 knots.
Approach speed (at max. landing wt.): about 130·knots.
Take-off distance to reach 50 ft., all engines operating (at max. t.o. wt.):
  At I.S.A. at sea level: 3,780 ft.
Landing distance from 50 ft. (at max. landing wt.): 3,800 ft.
Range (take-off to landing), still air, no reserves, I.S.A., max. fuel (5,213 imp. gal.): 3,020 n. mi. at 26,250 ft. (mean).

Range (take-off to landing), still air, no reserves, I.S.A., with max. payload: 1,350 n. mi.

Structure. Circular-section monocoque fuselage. Three-spar main wing section, 72-ft. span, and two-spar extension planes. Fail-safe construction.

Fuel system. Integral tanks between spars in outer planes, 20 bag tanks in inner wing. Capacity, 5,200 imp. gal. Four-point pressure refuelling.

Undercarriage. Four-wheel bogie main units, forwards retracting into inboard nacelles, and forwards-retracting twin-wheel steerable nosewheel. Hydraulic brakes. Track, 29 ft. 6 in. Base, 41 ft. 11 in. Tyre pressure, 113 p.s.i.

Flying controls. Manual, rod-operated, aerodynamically balanced ailerons, elevators and rudder; trim tabs on elevator and rudder, spring tab on rudder only. Autopilot.

Flaps, air brakes. Single-slotted flaps in one piece on each wing. No air brakes.

Cabin conditioning. Pressurization by engine-bleed air through heat exchangers. Differential, 7.1 p.s.i.

Hydraulics. 3,000-p.s.i. system with two NP-25 engine-driven pumps and two 2,100-p.s.i. stand-by nitrogen-charged accumulators for u/c, brakes, flaps, nosewheel steering and windscreen wipers.

Electrics. Eight 12-kW. engine-driven generators for 24-V. d.c. and four 8-kW. alternators and an invertor for single-phase a.c. and two invertors for 3-phase a.c.

De-icing. Electrothermal de-icing of wing, fin, tailplane, propellers, spinners and windscreen. Engine-bleed air for intake de-icing.

LOCKHEED ELECTRA (U.S.A.)

IN THE AUTUMN of 1954, design studies were made by Lockheed for a fairly small four-engined turboprop transport with a high-wing layout. Designed to meet a tentative American Airlines specification, this design was known as the Lockheed CL-303 and was projected with Rolls-Royce Dart or Napier Eland engines among others.

The requirement included a range of 750 miles and a cruising speed of 350 m.p.h. Further analysis of this design in relation to the potential market among U.S. domestic operators showed that it was too small to be economic in the hands of most airlines. There were also objections to the high-wing layout, because of doubts about its safety in wheels-up landings or sea ditchings.

American Airlines revised its specification, and the requirements of Eastern Air Lines also became known to Lockheed at this stage. A new project, the CL-310, was drawn with a low-wing layout and four Allison T-56 (510-D13) turboprop engines. To meet the American Airlines requirement, Lockheed provided in this design for a cruising speed of more than 400 m.p.h., and a range of 1,850 statute miles carrying an 18,000-lb. payload against a 50-knot headwind. The gross weight was 98,500 lb., and the span 95 ft.

Eastern Air Lines wanted a larger aeroplane than this, able to carry a 20,000-lb. payload for 2,500 miles against a 50-knot headwind at 450 m.p.h. After an interim attempt to meet this requirement with the CL-310 design carrying extra fuel in tip tanks, Lockheed scaled the whole project up, having gained American Airlines' agreement.

An increase of 100 sq. ft. in wing area (span up to 99 ft.) provided space for extra fuel as well as improving the take-off and landing characteristics, and Eastern accepted a range of 2,300 miles offered with this version of the project, which had a gross weight of 110,000 lb. Assured of orders from both the U.S. domestic operators, Lockheed decided to go ahead with the design in the summer of 1955, when it took the company designation of Model 188. The name Electra was chosen.

LOCKHEED **ELECTRA** (U.S.A.)

A Lockheed L-188A Electra in the markings of National Air Lines.

LOCKHEED ELECTRA (U.S.A.)

On June 8, 1955, an order for thirty-five Electras was announced from American Air-lines, followed by one for forty from Eastern Air Lines on September 27, 1955. Other orders followed in 1956 and 1957, principally from U.S. domestic airlines but including some useful business from Australasia and, alone among European airlines, KLM. By the time the first Electra flew, 144 were on order, and the total built eventually grew to 171.

From the time the decision was taken to proceed with the Electra to the scheduled introduction into service, a period of 44 months was allowed. In the course of this time, further development led to increases in the payload and gross weight. With small structural im-provements, these became, by September, 1957, 26,500 lb. and 113,000 lb. respectively.

After flight trials began, a second model, the L-188C, was developed with a gross weight of 116,000 lb. and features specifically for international operations. Both versions had optional provision in wing fillet tanks for an extra 750 imp. gallons of fuel.

The first Electra (N1881) flew on December 6, 1957; it was followed by the second (N1882) on February 13, 1958, and the fourth (N1884) on April 10, 1958. The third airframe was used for static proof loading tests before it flew. First of Eastern Air Lines' Electras, the fifth off the line, flew on May 19, 1958, and the sixth aircraft, which had been ordered by General Motors for powerplant development, flew in mid-June.

CAA certification of the L-188A was ob-tained on August 22, 1958. Deliveries to Eastern Air Lines did not begin until October, however, and a series of aircrew strikes then delayed its introduction into service until January 12, 1959. This date marked the first regular airline operation by a turbine-engined airliner of U.S. design and manufacture.

The first Electra for American was delivered in November, 1958, and went into service on January 23, 1959. Certification of the L-188C was obtained on July 10, 1959.

The full list of Electra customers—following several revisions of contracts while the aircraft were being built—is as follows, with the dates of initial service in parentheses:

American Airlines, thirty-five (January, 1959); Ansett-ANA, three (March, 1959);

Braniff, ten (July, 1959); Cathay Pacific, two (April, 1959); Eastern Air Lines, forty (Jan-uary, 1959); Federal Aviation Agency, one (January, 1961); Los Angeles Dodgers, one (1961); Garuda Empire Airways, three (1961); KLM, twelve (December, 1959); National Airlines, fifteen (April, 1959); Northwest, eighteen (August, 1959); Pacific Southwest, five (November, 1959); Qantas, four (Decem-ber, 1959); Sports Aloft, three (1962); TEAL, three (December, 1959); TAA, three (Nov-ember, 1959); Western, twelve (August, 1959).

One of the KLM Electras operated for a time on lease to Air Ceylon. One of the PSA Electras was leased, prior to its delivery, to the U.S. Navy for use at the Atlantic Missile Range, and that now operated as an executive transport for the Los Angeles Dodgers was previously used by General Motors. In 1961, American Airlines leased three of its Electras to REAL Aerovias in Brazil.

Two of the four Lockheed flight develop-ment aircraft went eventually to Cathay Pacific, one of them having earlier been used for a round-the-world demonstration flight in October and November, 1958. The first Electra of all was sold eventually to Pacific Southwest. Lockheed also succeeded in selling a version of the Electra to the U.S. Navy for anti-submarine duties, under the designation P3V-1 Orion. For a time, the third test Electra flew with aerodynamic mock-ups of the radomes carried by the Naval version.

In September, 1959, and again in March, 1960, an Electra crashed in circumstances which indicated structural failure of the wing in flight. Consequently, on March 25, 1960, a speed restriction was imposed, limiting the cruising speed to 275 knots I.A.S. at cruising height,

1. *Two prototypes of the Lockheed Electra, N1881 and N1882, both later delivered to airlines.*

2. *Eastern Air Lines' first Electra N5501.*

3. *An American Airlines' Electra, N6101A.*

4. *N9701C carrying Braniff International Airways' markings.*

5. *An L-188C, N130US, of Northwest Orient Airlines.*

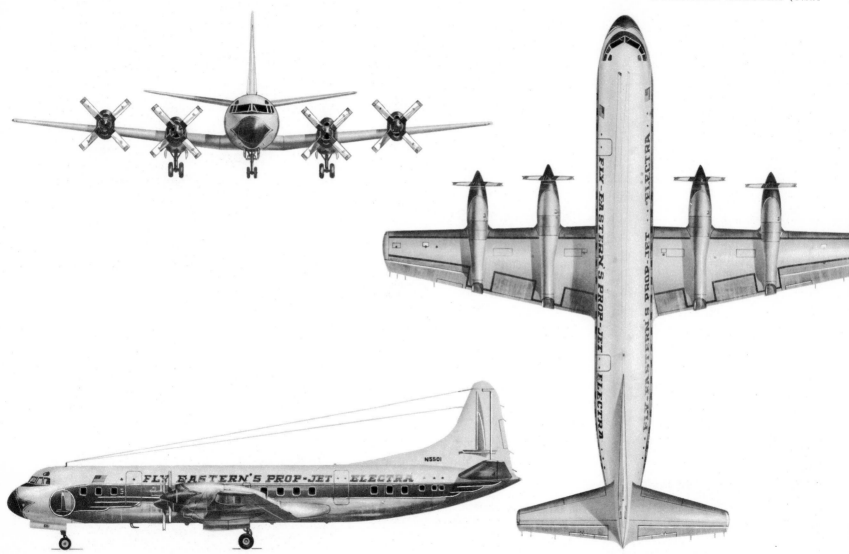

LOCKHEED ELECTRA (U.S.A.)

and this was reduced a few days later to 225 knots. This restriction was not lifted until December 31, 1960, after a painstakingly thorough investigation of the causes of these two accidents and tests of a series of modifications to prevent any recurrence.

The cause was pinned down to structural failure in turbulence of an airframe previously damaged as a result, for instance, of a heavy landing. If the powerplant structure had suffered slight damage in this way, of no significance in itself, it was possible for turbulence to set up an oscillation (whirl mode) of the engine and propeller which did not damp out but became catastrophic.

Modification of an Electra to have stronger structure in the susceptible area of the wing, and thicker gauge wing skins, began in July, 1960, and flight testing began in October. The FAA certificated the modifications on January 5, 1961, and Lockheed started work on a programme to modify all Electras delivered or awaiting delivery.

After modifications, the Electras were allowed to operate at their original cruising speeds and several airlines adopted designations such as Electra II to signify that the aircraft were up to the new standard. Eastern Air Lines was the first to resume unlimited operation of the Electra, on February 24, 1961.

6

7

LOCKHEED L-188C ELECTRA

Dimensions

Span: 99 ft. 0 in.
Overall length: 104 ft. 8 in.
Overall height: 32 ft. 10½ in.
Gross wing area: 1,300 sq. ft.
Sweepback: Nil.
Internal cabin dimensions:
    Length (ex. flight deck): 76 ft. 5 in.
    Max. width: 128 in.; max. height: 85 in.

Max. usable floor area (ex. flight deck): 704 sq. ft.
Max. usable volume (ex. flight deck): 4,812 cu. ft.

Accommodation

Typical mixed class: 26F (4 abreast at 38-in. pitch) and 52C (5 abreast at 34-in. pitch).
Max. high density: 104 (5 abreast at 34-in. pitch).
Volume of freight and baggage holds: 528 cu. ft.

6. *A Western Air Lines' L-188A Electra.*
7. *First of five Electras for PSA, N171PS.*
8. *The Qantas Electra VH-ECB Pacific Explorer.*

9. *In Ansett-ANA markings, L-188C VH-RMA.*
10. *A Trans-Australia Airlines' Electra VH-TLA.*

8

10

9

11. *The first L-188C Electra for KLM, PH-LLA.*
12. *TEAL Electra ZK-TEA Aotearoa.*
13. *The Cathay Pacific Electra VR-HFO.*
14. *One of three Garuda Indonesian Electras, PK-GLA.*
15. *A KLM L-188C Electra, PH-LLD, on lease to Air Ceylon.*

Powerplants

Four Allison 501D-13 turboprops.
Take-off power each (I.S.A., s.l.): 3,750 e.h.p. at 13,820 r.p.m.
Aeroproducts or Hamilton Standard four-blade 13-ft. 6-in. propellers.

Weights and loadings

Basic operational: 60,800 lb.
Total fuel: 43,700 lb.
Mfrs. max. payload: 26,300 lb.
Max. take-off: 116,000 lb.
Max. landing: 95,650 lb.
Max. zero fuel: 86,000 lb.
Wing loading (max. t-o. wt.): 89.2 lb./sq. ft.
Wing loading (max. landing wt.): 73.5 lb./sq. ft.
Power loading (max. t-o. wt.): 7.7 lb./e.h.p.

Performance

Typical cruising speed (I.S.A.): 353 knots at 20,000 ft. and 85,000 lb.; consumption: 553 imp. gal./hr.
Long-range cruise: 330 knots at 25,000 ft.
Approach speed ($1.3 V_{S_0}$ at max. landing wt.): 121 knots.
Take-off field lengths (FAA requirements at max. t-o. wt.):
At I.S.A. at sea level: 5,100 ft.
At I.S.A. +15° C. at sea level: 6,700 ft.
At I.S.A. at 5,000 ft.: 7,800 ft.
Landing field length (FAA requirements at max. landing wt.): 4,650 ft.
Range (take-off to landing), still air, no reserves, I.S.A., max. fuel (5,430 imp. gal.): 3,300 n. mi. with 11,500-lb. payload at 345 knots (mean) at 20,000 ft. (mean).

Range (take-off to landing), still air, no reserves, I.S.A., with max. payload: 2,360 n. mi. at 340 knots (mean) at 20,000 ft.

Structure. Semi-monocoque circular-section fuselage. Two-spar box-beam wing in two units and centre section integral with fuselage. Integrally stiffened wing panels are machined extrusions.

Fuel system. Two integral tanks between spars. Capacity, 4,596 imp. gal.; optional capacity, 5,429 imp. gal. Single-point pressure refuelling at 350 g.p.m.

Undercarriage. Menasco main and steerable nose oleos, backwards retracting. Goodrich wheels and brakes. Track, 31 ft. 2 in. Base, 37 ft. Tyre pressure, 114 p.s.i. at max. landing weight.

Flying controls. Hydraulically boosted ailerons, elevators and rudder, with manual bypass and manual trim. Bendix PB20E autopilot.

Flaps. Fowler-type flaps in one piece each side.

Cabin conditioning. Two engine-driven compressors. Freon vapour-cycle and two independent air-cycle cooling systems. Differential, 6.5 p.s.i.

Hydraulics. Two independent 3,000-p.s.i. systems with three electrically driven pumps for power controls, flaps, u/c, brakes and nosewheel steering.

Electrics. Four engine-driven 60-kVA. generators for 120/208-V. 400-cycle 3-phase a.c., and two rectifiers for 28-V. d.c.; 24-V. battery.

De-icing. Airframe de-icing by engine-bleed air. Electric de-icing for windscreen and propellers.

NAMC **YS-11** (Japan)

WORK ON A medium-range turboprop transport of local design began in Japan in 1957, on the initiative of the Japanese Ministry of International Trade and Industry. The task of designing and building such an aeroplane was more ambitious than anything else attempted in Japan since the end of World War II. A consortium of manufacturers was therefore formed, with the six major aircraft companies represented.

In May, 1957, these six companies set up the Transport Aircraft Development Association, to co-ordinate the design work which was already being undertaken on the new project. This Association was responsible for establishing the basic specification of the YS-11, as the project became designated, and for the construction of a full-scale mock-up.

To continue development of the YS-11 and undertake construction of prototypes, the Nihon Aeroplane Manufacturing Co. was formed in June, 1959, with the six original manufacturers and the Government sharing in financing. The six companies involved are: Mitsubishi, Kawasaki, Fuji, Shin Meiwa, Nihon Hikoki (Japan Aircraft), and Showa Hikoki.

Nihon Aeroplane Manufacturing Co., as the co-ordinating agency, placed contracts with the six companies to cover construction of four prototypes, of which two are intended for static and fatigue testing. The first flying prototype was completed in the spring of 1962, and further contracts had by then been placed by the Government for jigs, materials and components for production of the YS-11. Production models are to be available by the end of 1963, and both Japan Air Lines and All Nippon Airways have announced plans to purchase and operate the type.

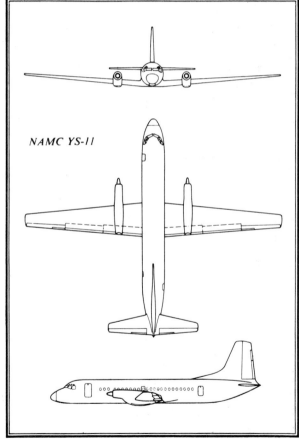

NAMC YS-11

NIHON YS-11

Dimensions

Span: 105 ft. 0 in.
Overall length: 86 ft. 3 in.
Overall height: 30 ft. 0 in.
Gross wing area: 1,020 sq. ft.
Sweepback: Nil.
Internal cabin dimensions:
  Length (ex. flight deck): 44 ft. 1 in.
  Max. width: 106 in.; max. height: 78 in.
Max. usable floor area (ex. flight deck): 340 sq. ft.
Max. usable volume (ex. flight deck): 2,125 cu. ft.

Accommodation

Normal tourist: 54 (4 abreast at 34-in. pitch).
Max. high density: 60 (4 abreast at 34-in. pitch).
Volume of freight and baggage holds: 67 cu. ft.

Powerplants

Two Rolls-Royce Dart 542 (R.Da.10/1).
Take-off power each (I.S.A., s.l.): 3,060 e.h.p. at 15,000 r.p.m.
Water methanol injection. Rotol four-blade 14-ft. 6-in. reversing propellers.

Weights and loadings

Basic operational: 31,217 lb.
Total fuel: 12,600 lb.
Mfrs. max. payload: 12,125 lb.
Max. take-off: 50,265 lb.
Max. landing: 48,050 lb.
Max. zero fuel: 44,100 lb.
Wing loading (max. t-o. wt.): 49.4 lb./sq. ft.
Wing loading (max. landing wt.): 47.1 lb./sq. ft.
Power loading (max. t-o. wt.): 8.2 lb./e.h.p.

Performance

Typical cruising speed (I.S.A.): 253 knots at 20,000 ft. and 44,092 lb.; consumption: 6.6 lb./n. mi.
Approach speed ($1.3 V_{s_0}$ at max. landing wt.): 94 knots with 40-deg. flap.
Take-off field lengths (FAA requirements at max. t-o. wt.):
  At I.S.A. at sea level: 2,890 ft.
  At I.S.A. +23° C. at sea level: 3,200 ft.
  At I.S.A. at 6,000 ft.: 4,080 ft.
Landing field length (FAA requirements at max. landing wt.): 3,790 ft.
Range (take-off to landing), still air, no reserves, I.S.A., max. fuel (1,560 imp. gal.): 1,680 n. mi. with 5,400-lb. payload at 247 knots (mean) at 20,000 ft. (mean).
Range (take-off to landing) still air, no reserves, I.S.A., with max. payload: 700 n. mi.

Structure. Circular-section semi-monocoque fuselage. Straight-tapered torsion-box wing in two sections joining centre section through fuselage. Fail-safe construction.

Fuel system. Integral tanks in wings outboard of nacelles. Capacity, 1,005 imp. gal.; optional bag tanks inboard of nacelles, capacity 490 imp. gal. Water-methanol tank in port wing, capacity 105 imp. gal.

[continued on p. 85

NORD SUPER BROUSSARD (France)

DEVELOPED BY Avions Max Holste (now Rheims Aviation) on the basis of experience with the single-engined Broussard utility aircraft, the Super Broussard is now being marketed by Nord Aviation, one of the French state factories. The Super Broussard project began in 1957, as a design for a small transport and feeder-liner with two radial engines.

Designated MH-250, a prototype of this project was built by Max Holste, and flew for the first time on May 20, 1959. The engines were Pratt & Whitney R-1340s. A turboprop variant had also been projected as the MH-260, and this replaced the MH-250 as the basic version of the series.

Powered by two Turboméca Bastan IIIA turboprops, an MH-260 prototype (F-WJDV) made its first flight on July 29, 1960. Apart from the new powerplant, the MH-260 differed from the MH-250 primarily in fuselage length, which was 4 ft. 7 in. longer. Bastan IVs were fitted in the prototype in October, 1960, and these are the production standard powerplant.

On November 23, 1960, an agreement between Nord and Max Holste gave the former company manufacturing and marketing rights, and Government finance was obtained to initiate Nord production. The first aircraft off this line (F-WJSN) flew at Toulouse on January 29, 1962.

In parallel with the MH-260, Nord has developed the MH-262, which has a pressurized, circular-section fuselage, and can accommodate up to 29 passengers. A prototype was to fly in the autumn of 1962, and production aircraft were offered to airlines by the second quarter of 1963.

The MH-262 is powered by 1,065-e.h.p. Bastan VI engines and has a rather better performance than its unpressurized counterpart in most respects, but carries a smaller payload.

ABOVE—*Prototype Max Holste MH-260 Super Broussard in Royal Air Maroc colours.*

NORD MH-260 SUPER BROUSSARD

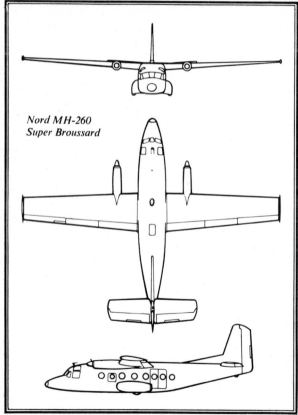

Nord MH-260 Super Broussard

Dimensions

Span: 71 ft. 10 in.
Overall length: 59 ft. 2 in. (MH-262, 63 ft. 3 in.)
Overall height: 23 ft. 4 in. (MH-262, 21 ft. 8 in.)
Gross wing area: 592 sq. ft.
Sweepback: Nil.
Internal cabin dimensions:
   Length (ex. flight deck): 24 ft. 5 in.
   Max. width: 86 in.; max. height: 71 in.
Max. usable floor area (ex. flight deck): 169 sq. ft.
Max. usable volume (ex. flight deck): 1,193 cu. ft.

Accommodation

Typical tourist: 20 (3 abreast at 35-in. pitch).
Max. high density: 23 (3 abreast at 30-in. pitch).

Powerplants

Two Turboméca Bastan IV turboprops.
Take-off power each (I.S.A., s.l.): 986 e.h.p.
Ratier Figeac FH86 three-bladed 9-ft. 6-in. propellers.

Weights and loadings

Basic operational: 12,405 lb. (MH-262, 12,933 lb.).
Total fuel: 3,650 lb.
Mfrs. max. payload: 8,098 lb. (MH-262, 7,570 lb.).
Max. take-off: 21,605 lb. (MH-262, 21,715 lb.).

Max. landing: 20,720 lb.
Max. zero fuel: 20,500 lb.
Wing loading (max. t-o. wt.): 36.4 lb./sq. ft.
Wing loading (max. landing wt.): 35.0 lb./sq. ft.
Power loading (max. t-o. wt.): 11.0 lb./e.h.p.

Performance

Best cost. cruising speed (I.S.A.): 187 knots (MH-262, 205 knots) at 10,000 ft. and 21,380 lb.; consumption: 90 imp. gal./hr.
High-speed cruise: 200 knots (MH-262, 217 knots).
Approach speed (at max. landing wt.): 85 knots.
Take-off field lengths (FAA requirements at max. t-o. wt.):
   At I.S.A. at sea level: 3,900 ft.
Landing field length (FAA requirements at max. landing wt.): 3,715 ft.
Range (take-off to landing), still air, no reserves, I.S.A., max. fuel (455 imp. gal.): 850 n. mi. (MH-262, 980 n. mi.) with 6,200-lb. payload at 187 knots (mean) at 10,000 ft. (mean).
Range (take-off to landing), still air, no reserves, I.S.A., with max. payload: 202 n. mi. at 187 knots at 10,000 ft.

Structure. Semi-monocoque fuselage of roughly rectangular cross-section with curved sides.

NORD SUPER BROUSSARD (France)

Torsion box wing in two halves attached to separate centre section. Fabric covered control surfaces.

Fuel system. Six light-alloy tanks in wings; gravity refuelling. Capacity, 396 imp. gal.

Undercarriage. Semi-independent main oleo pneumatic shock absorbers retract backwards into fuselage-side housings: Oleo pneumatic self-stabilizing nosewheel. Hydraulic brakes. Track, 9 ft. 10 in. Base, 18 ft. Tyre pressure, 54 p.s.i.

Flying controls. All manual system, cable operated. Aerodynamically balanced ailerons. elevators and rudder.

Flaps, air brakes. Slotted flaps in two sections each side. No air brakes.

Cabin conditioning. Unpressurized cabin with air conditioning by individual ventilators and glass wool sound-proofing.

Hydraulics. Hydraulic operation of flaps, undercarriage and brakes.

Electrics. Two 6-kW. engine-driven generators and a 28-V. 36-amp. hr. storage battery in the fuselage nose.

De-icing. Pneumatic wing and tail de-icing. Liquid protection of propellers.

(*continued from p. 83*)

NAMC YS-11 (Japan)

Undercarriage. Forwards retracting main and nosewheel oleos with low-pressure tyres. Brakes and anti-skid units. Track, 28 ft. 3 in. Base, 30 ft. 11 in.

Flying controls. Fully manual system, with tabs on all control surfaces. Space provision for autopilot.

Flaps, air brakes. Fowler-type flaps in two sections each side. No air brakes.

Cabin conditioning. Four engine-driven superchargers, including one stand-by to provide pressurization at differential of 4.16 p.s.i. Ground conditioning from one engine.

Hydraulics. 3,000-p.s.i. system for flaps, u/c, nosewheel steering, brakes, propeller brakes, windscreen wipers and ground cooling blower.

Electrics. Two 12-kW. engine-driven generators and two 24-amp. hr. batteries for 28-V. d.c. Two 25-kVA. and one 0.5-kVA. invertors and four 30-kVA. "frequency-wild" alternators.

De-icing. Hot air wing and tail de-icing using combustion heaters. Electric de-icing of windscreen, propellers and intakes.

SUD-AVIATION CARAVELLE (France)

ALTHOUGH THE MAJORITY of new jet transport projects since 1960 have the engines located at the rear of the fuselage, this layout was a novelty when it was adopted for the Caravelle. The French design team under Pierre Satre which was responsible for this original but highly successful aircraft was not, in fact, the first to study this arrangement. Among other companies, de Havilland had explored similar layouts when the Comet was in the preliminary design stage. The Caravelle was, however, the first aircraft actually built with engines at the rear.

The Caravelle design began when the official French civil aviation agency, the SGACC, drew up a basic specification for a medium-range airliner. This was in 1951, and the aircraft was to carry a payload of 6–7 tons for 1,000–1,200 miles at better than 330 knots. The French industry was still recovering from World War II and this specification was a positive step by the Government to encourage development of an aircraft likely to have a world sales appeal without competing head-on with American long-range transport aircraft.

Preliminary design studies were submitted by a number of companies, both private and state-owned, and ranged from piston-engined through turboprop and turbojet types to some using ducted fans. Two-, three- and four-engined projects were submitted.

The design competition was won by the nationalized Sud-Est (SNCASE) company, which merged in 1957 with Sud-Ouest (SNCASO) to form the present Sud-Aviation. The SNCASE design team had been projecting commercial airliners of various types since 1946 and had studied about fifty designs by 1951. These carried project designations from X200 to X210 and included turboprop types as well as turbojets with podded, buried and fuselage-mounted engines.

The X210 designation itself covered a number of designs, but by the end of 1951 the design team had come to the view that a pure-jet type would best meet the SGACC specification, because of its high performance, basic simplicity and easy maintenance. Only one suitable engine was available in France, the SNECMA Atar, which was in production for the French Air Force.

For the aircraft weight already decided upon, three of these engines were needed, and this led inevitably to consideration of the rear fuselage for their location, with one inside the structure and one each side in external nacelles. A good deal of design and wind tunnel work was done on this "Tri-Atar" project.

Further consideration of the choice of powerplant led to a decision that the Rolls-Royce Avon held greater potential for development and a better background of service. As the Avon was more powerful than the Atar, only two would be required, and the X210 was reconsidered with the centre engine deleted and one Avon in each nacelle.

Various design studies were submitted for evaluation in January, 1952, and the twin-Avon X210 was confirmed as the winner in September the same year. An order was placed for four prototypes, including two for static and fatigue tests and, in due course, the French Government gave financial guarantees to allow full-scale production to proceed in the Sud-Est factory at Toulouse.

The prototype contract was signed on January 3, 1953, and the two flying prototypes, with Avon R.A.26 engines, made their first flights on May 25, 1955 (F-WHHH) and May 6, 1956 (F-WHHI). The first aircraft completed its maker's test programme and the French certification trials by April 30, 1956, and was then handed over to Air France, which had placed the first order for Caravelles in November 16, 1955 (for twelve plus an option on twelve more which was taken up on July 10, 1958). On these trials, 173 flights were made totalling 411 flight hours.

F-BHHH was then handed over to Air France for route proving. Between May 23 and October 2, 1956, it made 209 flights totalling 505 hours, principally between Paris and Algiers. The second prototype, which was the first with a furnished cabin, was used between April 18 and June 25 for an extensive tour of South and North America.

Few alterations were made from the prototypes in the production model, the first of which flew on May 18, 1958. This was a Caravelle I; others up to Caravelle XIV have since appeared and are noted below.

SUD-AVIATION **CARAVELLE** (France)

The prototype Caravelle VI-R, F-WJAP, in United Air Lines markings.

SUD-AVIATION CARAVELLE (France)

Caravelle I. The first production batch of Caravelles were of this type, with 10,760-lb.-thrust Avon R.A.29 Mk. 522 engines. Compared with the prototypes, the production model has a long, slender extension of the dorsal fin along the back of the fuselage to contain aerials; a fuselage longer by 4.6 ft. and weather radar. Gross weight was increased to 95,900 lb.

Ten of the Air France aircraft ordered in February, 1956, were delivered to this standard, the first on March 19, 1959. Regular services began on May 12, 1959. The second Caravelle customer was SAS, which took delivery of the first of six Caravelle Is on April 10, 1959, and put the type into service on May 15. The Brazilian airline Varig took delivery of two Caravelle Is in 1959 and Air Algérie received the only other two of this type made. In 1962, Air Algérie disposed of its two Caravelle Is, one going to Varig as a replacement for one lost in 1961, and the other going to Air Liban.

Caravelle IA. Minor changes in the Avon R.A.29/1 engine, to produce the Mk. 526, changed the designation of the later Caravelle Is. There were no external differences. Air France received two Caravelle IAs; SAS six, Finnair two, Air Algérie one, and Royal Air Maroc one.

Caravelle III. Initially, Caravelle designations were linked with the stage of Avon engine development. Thus, while the Caravelle I had the R.A.29 Stage 1, the Caravelle III had Stage 3 engines and the Caravelle VI had R.A.29 Stage 6s; the designations II, IV and V were not used. With improved turbine blade material, the Avon R.A.29/3 produced 11,700-lb. static thrust. Coupled with the results of static testing of the fourth Caravelle airframe at Toulouse, this made possible an increase in gross weight to 101,400 lb., and in cruising speed from 405 knots to 430 knots.

Air France increased its Caravelle order by twelve on July 10, 1958 and additional orders since then have brought the total to forty, plus two more ordered on behalf of Air Liban and Tunisair. All but the first twelve of these are Caravelle IIIs, and the earlier aircraft have

been modified to III standard. Other Series IIIs have been delivered to SAS (three, plus four leased to Swissair), Finnair (one), Air Algerie (one), Alitalia (four), and Royal Air Maroc (one).

Two other Caravelle IIIs were built—one as a Sud prototype (F-WJAQ), first flown on December 30, 1959 and later modified to the prototype VI and then sold to Aerolineas. The other was built for General Electric to serve as a prototype for the Caravelle VII.

Caravelle VI. With an extra compressor stage, the Avon R.A.29-6 developed 12,500-lb. thrust and allowed further weight and performance growth of the Caravelle. In the Caravelle VI, the weight increased to 105,800 lb. and the cruising speed to 445 knots.

The new Avons, Mk. 531, were first fitted on the Caravelle III prototype (F-WJAQ) which first flew as a Caravelle VI on September 10, 1960. A 352-hour flight programme (in 235 flights) was completed on this aircraft before deliveries began in January, 1961, to Sabena.

Sabena ordered eight Caravelle VIs. Other orders for this variant—which is sometimes called the VIN to distinguish it from the VIR—were from Alitalia (for 10); Air Algérie (for two); Aerolineas Argentinas (three) and Jugoslovenskie Aero Transport (three).

Caravelle VIR. Development of the Caravelle particularly to meet American requirements produced this version, in which the "R" indicates the reverse-thrust engines. With this feature, the Avons are designated Mk. 532R or 533R, with a rating of 12,920 lb.s.t. Several other new features were introduced to the requirements of United Air Lines, including large windows in the cockpit, additional

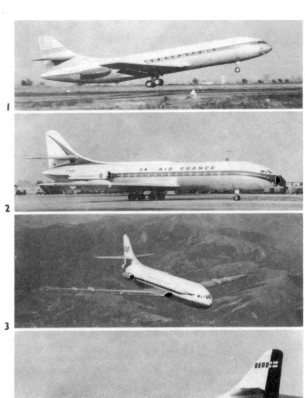

1. *The first Caravelle prototype, F-WHHH, with short dorsal fin.*
2. *A production Caravelle III, F-BHRU, in Air France markings.*
3. *The first prototype, re-registered as F-BHHH and modified to production standard with the dorsal spline, in SAS markings.*
4. *A Finnair Caravelle IA.*
5. *Varig's second Caravelle IA, PP-VJD.*

SUD-AVIATION **CARAVELLE** (France)

SUD-AVIATION CARAVELLE (France)

spoilers on the wing to improve the take-off, more powerful brakes to shorten the landing and some other details.

Sud built a prototype of this model (F-WJAP), which first flew, in United colours, on February 6, 1961. The United order for twenty had been announced in February, 1960, and deliveries began in June, 1961, for United to fly their first service on July 14. Other orders for the Caravelle VIR were placed by Iberia (six), Panair do Brasil (four) and TAP (three).

Caravelle VII. Originally applied to a Caravelle VI with lengthened fuselage this designation was later switched to the version with General Electric CJ805–23C turbofans. The U.S. engine company purchased a Caravelle III, named **Santa Maria**, for demonstrations and after making a U.S. tour this aircraft was fitted with the new engines and flew for the first time on December 29, 1960.

Production models were planned with these engines and a fuselage 1 metre (3 ft. 4 in.) longer, but the designation was changed to Caravelle 10.

Caravelle VIII. This designation covered a projected long-fuselage Caravelle with R.B.141-3 engines.

Caravelle 10. To meet TWA requirements, Sud developed the longer Caravelle with General Electric engines, and adopted the designation Caravelle 10A in place of Caravelle VII. New features of this model were an increase of wing chord at the leading edge near the root; lowering of the window line; introduction of an "acorn" at the fin/tailplane junction; use of double-slotted flaps and other improvements.

TWA ordered twenty Caravelle 10As in 1961 but was unable to finance the order, which was cancelled in 1962.

Other versions of the Caravelle 10 projected by Sud were the Caravelle 10B with Pratt & Whitney JT8D-1 turbofans; the Caravelle 10C, with a high-density layout for ninety-nine passengers and Avon 533 engines; and the Caravelle 10D, with Rolls-Royce R.B.174 engines.

Caravelle 14. This projected variant was the outcome of a sales and technical agreement between Sud and Douglas concluded in December, 1959, making use of Douglas experience in wing design for the DC-8. The engines were to be 15,000-lb. R.B.141-11A turbofans.

---

6. *Swissair's Caravelle I in pre-delivery registration, F-WJAM.*
7. *The Royal Air Maroc Caravelle I registered as F-WJAL before delivery.*
8. *Air Algérie's Caravelle III F-OBNG.*
9. *The first Caravelle VI, OO-SRA for Sabena.*
10. *Alitalia Caravelle VI, I-DAXA.*

SUD-AVIATION CARAVELLE

Dimensions

Span: 112 ft. 6 in.
Overall length: 105 ft. 0 in.
Overall height: 28 ft. 7 in.
Gross wing area: 1,579 sq. ft.
Sweepback: 20 deg. on the quarter-chord line.
Internal cabin dimensions:
    Length (ex. flight deck): 72 ft. 0 in.
    Max. width: 118.5 in.; max. height: 82 in.
Max. usable floor area (ex. flight deck): 646 sq. ft.
Max. usable volume (ex. flight deck): 4,200 cu. ft.

Accommodation

Typical mixed class: 20F (4 abreast at 37-in. pitch) and 52 (5 abreast at 37-in. pitch).
Max. high density: 80 (5 abreast at 37-in. pitch).
Volume of freight and baggage holds: 534 cu. ft.

CARAVELLE III

Powerplants

Two Rolls-Royce Avon 527 turbojets.
Take-off power each (I.S.A., s.l.): 11,400 lb. at 8,000 r.p.m.
No water injection. No thrust reversers. Sound suppressors fitted.

Weights and loadings

Basic operational: 59,686 lb.
Total fuel: 33,500 lb.
Mfrs. max. payload: 18,300 lb.
Max. take-off: 101,413 lb.
Max. landing: 96,584 lb.
Max. zero fuel: 78,264 lb.
Wing loading (max. t-o. wt.): 64.23 lb./sq. ft.
Wing loading (max. landing wt.): 61.16 lb./sq. ft.
Power loading (max. t-o. wt.): 4.45 lb./lb. thrust.

Performance

Best cost. cruising speed (I.S.A.): 414 knots at
    35,000 ft. and 82,000 lb.; consumption: 680
    imp. gal./hr.
High-speed cruise: 432 knots at 25,000 ft.
Long-range cruise: 396 knots at 35,000 ft.
Approach speed (at max. landing wt.): 123 knots.
Take-off field lengths (FAA requirements at max.
    t-o. wt.):
        At I.S.A. at sea level: 6,100 ft.
        At I.S.A.+15° C. at sea level: 7,000 ft.
        At I.S.A. at 5,000 ft.: 9,100 ft.
Landing field length (FAA requirements at max.
    landing wt.): 5,900 ft.
Range (take-off to landing), still air, no reserves,
    I.S.A., max. fuel (4,163 imp. gal.): 2,315 n. mi.
    with 8,430-lb. payload at 400 knots (mean) at
    35,000 ft. (mean).
Range (take-off to landing), still air, no reserves,
    I.S.A., with max. payload: 1,565 n. mi. at 390
    knots (mean) at 35,000 ft. (mean).

CARAVELLE VI-R

Powerplants

Two Rolls-Royce Avon 532R turbojets.
Take-off power each (I.S.A., s.l.): 12,600 lb. at
    8,150 r.p.m.
No water injection. Thrust reversers and sound
    suppressors fitted.

Weights and loadings

Basic operational: 63,856 lb.
Total fuel: 33,500 lb.
Mfrs. max. payload: 17,715 lb.
Max. take-off: 110,232 lb.

---

11. *Air Liban Caravelle III OD-ADY.*
12. *Tunis Air Caravelle III TS-IKM.*
13. *The Caravelle VII demonstrator N420GE*
    Santa Maria.
14. *United Air Lines' N1006U, a Caravelle
    VI-R.*

Max. landing: 104,980 lb.
Max. zero fuel: 79,400 lb.
Wing loading (max. t-o. wt.): 69.81 lb./sq. ft.
Wing loading (max. landing wt.): 66.49 lb./sq. ft.
Power loading (max. t-o. wt.): 4.37 lb./lb. thrust.

Performance

Best cost. cruising speed (I.S.A.): 423 knots at
    35,000 ft. and 91,000 lb.; consumption: 780
    imp. gal./hr.
High-speed cruise: 456 knots at 25,000 ft.
Long-range cruise: 411 knots at 35,000 ft.
Approach speed (at max. landing wt.): 128 knots.
Take-off field lengths (FAA requirements at max.
    t-o. wt.):
        At I.S.A. at sea level: 6,650 ft.
        At I.S.A.+15° C. at sea level: 7,400 ft.
        At I.S.A. at 5,000 ft.: 9,400 ft.
Landing field length (FAA requirements at max.
    landing wt.): 5,415 ft.
Range (take-off to landing), still air, no reserves,
    I.S.A., max. fuel (4,163 imp. gal.): 2,085 n. mi.
    with 13,195-lb. payload at 409 knots (mean) at
    35,000 ft. (mean).
Range (take-off to landing), still air, no reserves,
    I.S.A., with max. payload: 1,780 n. mi. at
    403 knots (mean) at 35,000 ft. (mean).
Structure. Circular-section monocoque fuselage.
    Torsion-box wing structure with three spars,
    built in two pieces joined on fuselage centre line.
Fuel system. Two integral tanks in each wing.
    Capacity, 4,160 imp. gal. Two-point pressure
    refuelling at 500 g.p.m.
Undercarriage. Four-wheel bogie main units and
    steerable twin nosewheel. Track, 17 ft. Base,
    38 ft. 7 in. Tyre pressure, 124 p.s.i.
Flying controls. Hydraulically powered two-piece
    ailerons, rudder and elevator, with artificial
    feel. Lear L-102 autopilot.
Flaps, air brakes. Slotted flaps (double slotted for
    Series 10) in four sections each side. Spoilers
    above and below each wing ahead of flaps and
    additional spoilers above wing on Caravelle
    VI-R.
Cabin conditioning. Two compressors driven by
    engine-bleed air. Differential, 8.05 p.s.i.
Hydraulics. Twin 2,500-p.s.i. systems (each with
    electrically powered stand-by) for flying con-
    trols, u/c, wheel brakes, flaps, spoilers, nose-
    wheel steering, airstairs and feel simulators.
Electrics. Two engine-driven 30-V. d.c. generators
    provide 28.5-V. d.c. and 115-V. 400-cycle a.c.
    through two invertors. 26-V. a.c. available also
    through transformer.
De-icing. Wings, tail unit and intakes by engine-
    bleed air.

TUPOLEV Tu-104 (U.S.S.R.)

Tupolev Tu-104B, CCCP-42430, landing at Le Bourget.

THE SECOND JET transport in regular air-line service, after the de Havilland Comet, was Russia's Tupolev Tu-104. Operating on the route Moscow–Irkutsk on September 15, 1956, it preceded the first service by an American turbojet type by more than two years.

Development of the Tu-104 had begun in or about 1953 and was afforded high priority in order to provide Russia's airline system with a prestige vehicle. Operating economy and, to a lesser extent, passenger comfort were subordinate to this aim. The first Tu-104 (believed to be CCCP-L5400) made its first flight early in 1955, followed the same year by others from a pre-production batch.

Only a year later, the type made its first overseas appearance with a flight to London in March, 1956. Proving and demonstration flights to other European destinations, to China and to many cities in U.S.S.R. followed.

The first batch of Tu-104s had a basic layout with three cabins in the forward fuselage seating twenty-two passengers in all, and twenty-eight more in the rear cabin. Between the two sections was a small galley, located over the wing centre section where it passed through the fuselage. This interruption of cabin space by the basic wing structure was a reminder that the Tu-104 design was derived with minimum modification from the Tu-16 twin-jet bomber. Only the fuselage was new.

In mid-1957, a new version called the Tu-104A was demonstrated at Vnukovo. Said to have "improved" engines, it also had a revised layout with accommodation for seventy passengers, including fifty-four in the rear cabin. This version became the most used by Aeroflot and in 1962 its further development into a high-density variant was announced. By rearranging the galleys and toilets, seats for one hundred are provided.

A 100-seat version known as the Tu-104B had previously gone into service—in 1959—
[continued opposite

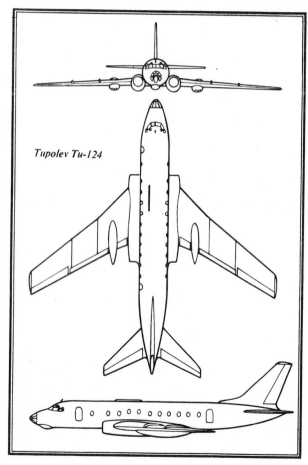

Tupolev Tu-124

ABOVE—*Tupolev Tu-104A OK-LDB, one of four supplied to CSA in Czechoslovakia.*

TUPOLEV Tu-104

Dimensions
Span: 113 ft. 4 in.
Overall length: 127 ft. 6 in.
Overall height: 39 ft. 6 in.
Sweepback: 36 deg. on quarter-chord line.
Internal cabin dimensions:
    Max. width: 126 in.; max. height: 83 in.
Max. usable volume (ex. flight deck): 5,650 cu. ft.

Accommodation
Typical mixed class: 16F (4 abreast) and 54E (5 abreast).
Max. high density: 100 (5 abreast).
Volume of freight and baggage holds: 530 cu. ft.

Powerplants
Two Mikulin AM-3 turbojets.
Take-off power each (I.S.A., s.l.): 19,180 lb.

Weights and loadings
Total fuel: 60,000 lb.
Mfrs. max. payload: 29,430 lb.
Max. take-off: 166,450 lb.
Max. landing: 141,100 lb.
Wing loading (max. t-o. wt.): 91.2 lb./sq. ft. (estimated).
Power loading (max. t-o. wt.): 3.7 lb./lb. thrust.

Performance
High-speed cruise: 513 knots at 27,000 ft.
Long-range cruise: 432 knots at 30,000 ft.
Approach speed (at max. landing wt.): 144 knots (estimated).
Take-off distance to 50 ft. (at max. t.o. wt.):
    At I.S.A. at sea level: 5,413 ft.
Landing distance from 50 ft. (at max. landing wt.): 5,052 ft.
Range (take-off to landing), still air, no reserves, I.S.A., max. fuel (7,292 imp. gal.): 2,266 n. mi. with 17,637-lb. payload.

Structure. Circular-section monocoque fuselage. Two-spar wing in two main units joined by a fuselage-width centre section.

Fuel system. Three bag tanks in each wing and one in centre section. Capacity, 7,292 imp. gal. Gravity refuelling.

Undercarriage. Four-wheel bogie main units, retracting up and back into wing nacelles; twin-wheel steerable nosewheel. Braking parachute fitted.

Flying controls. Fully manual ailerons, elevator and rudder.

Flaps, air brakes. Fowler-type flaps in two portions on each wing. No air brakes.

Cabin conditioning. Differential, 7.2 p.s.i.

De-icing. Thermal de-icing of wing leading edge by engine-bleed air.

TUPOLEV Tu-124 (U.S.S.R.)

PHOTOGRAPHS OF THE Tupolev Tu-124 were first published in the Soviet Press during 1960, probably soon after the first flight of this short-haul jet transport. In appearance, the type has an obvious family resemblance to the Tu-104, Russia's first jet transport, and the fuselage cross section may be the same, as well as the wing and much of the tail unit.

The fuselage is shorter than that of the Tu-104, and the Tu-124 is reported to provide accommodation for forty-four passengers in the basic version or up to sixty-eight in high-density layouts. The most interesting aspect of the Tu-124 is that it is powered by Soloviev turbofan engines. It is thus the first Russian airliner to make use of this new type of engine, and precedes by about two years the BAC One-Eleven which is in a similar size and performance bracket.

The Tu-124 was expected in service before the end of 1962 on the more important short Aeroflot routes, such as Moscow–Perm (725

TUPOLEV Tu-104 *continued*]

on the Moscow–Leningrad route and elsewhere within the Soviet Union. This version, however, had a slightly longer fuselage.

Although Russia announced that the Tu-104 was available for export and put some effort into promoting overseas sales, only CSA has purchased the type. The four Czech Tu-104As are OK-LDA *Praha*, OK-LDB *Bratislava*, OK-LDC *Brno*, and OK-NDD *Plzen*.

miles) and Moscow–Chelyabinsk (925 miles). First illustrations showed an aircraft numbered CCCP-L45000 but early in 1962 CCCP-L45112 was illustrated, suggesting that deliveries were already being made in some numbers.

Power of the turbofan engines is reported to be 8,500 lb. Few detailed figures for the Tu-124 are available but it is said to cruise at between 500 and 560 m.p.h. at heights up to 40,000 ft., and to have a maximum range of 1,240 miles. It can operate from airfields having runways of less than 6,000 ft. in length.

A flight crew of three—two pilots and an engineer—operate the Tu-124, and one stewardess is usually carried. The cabin is divided into three sections, with a step up over the main wing spar where it passes through the fuselage, as in the Tu-104. Two main doors are provided, one forward and one aft of the wing.

Early in 1962, there were reports from Russia of a Tu-124A which has two engines mounted Caravelle-fashion on the rear fuselage. As the engine installation in the wing roots of the Tu-124 is a major structural feature, the Tu-124A must be more than a simple second variant of the basic design. No illustrations had been published of this version up to the spring of 1962.

1. *An early production model Tu-124, CCCP-45005, showing its similarity with the Tu-104.*

2. *CCCP-45000, believed to be the first example of the Tu-124 built.*

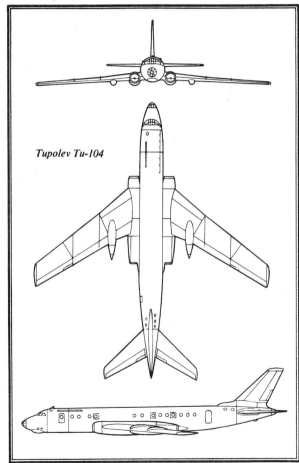

Tupolev Tu-104

2

TUPOLEV Tu-114 (U.S.S.R.)

HAVING SUCCESSFULLY CONVERTED the Tu-16 bomber into a jet transport as the Tu-104 (see page 92), the Tupolev design bureau set about a similar modification of the larger Tu-20. The work was programmed to produce the new transport—the world's largest—to coincide with the fortieth anniversary of the October Revolution in 1957.

The first flight of the prototype was probably made towards the end of that year. Photographs of it in the air first appeared in February, 1958. Like the Tu-104, the Tu-114 had a new fuselage while retaining the wings, tail unit and powerplant of its military counterpart. The fuselage, over 150 ft. long, has a diameter of 12 ft. 4 in., making it the only transport to date able to accommodate eight seats abreast.

Maximum accommodation in the Tu-114 has been reported to be 220, but 170 seems to be the normal standard on domestic flights and 120 for international flights. A cabin amidships is reserved for use as a "restaurant" and is not normally occupied during take-off. The galley space includes a kitchen below floor level in the wing centre section.

1. *A Tupolev Tu-114D, CCCP-76462, with the small fuselage which distinguishes this variant.*
2 and 3. *The Tupolev Tu-114 prototype.*
4. *The Tu-114 CCCP-L5611 at Paris.*

For some time after the Tu-114 first appeared, little more was heard of it, although Russia announced that it was to be produced for Aeroflot. Plans to start service were twice announced and then postponed, scheduled operations eventually starting on April 24, 1961, on the route between Moscow and Khabarovsk. This 4,350-mile sector is flown non-stop in about 8½ hours.

Prior to this, the Tu-114 had made several overseas demonstration flights, of which the first was to Paris in June, 1959, for the International Air Show there. Later the same month, the Tu-114 flew from Moscow to New York non-stop.

On March 24, 1960, a Tu-114 made a record-breaking flight carrying a payload of 25,000 kg. over a 1,000-km. (620-mile) circuit at a speed of 541.45 m.p.h. This performance represented records in eight FAI categories, for speed over the 1,000-km. distance with nil, 1,000-kg., 2,000-kg., 5,000-kg., 10,000-kg., 15,000-kg., 20,000-kg. and 25,000-kg. payloads.

This was the first of three such flights. The second was on April 1, 1960, when the same payload was carried over a 2,000-km. (1,240-mile) distance at 532.7 m.p.h. This set eight more records and a third series of eight records was set on April 9, 1960, when the 25,000-kg. payload was carried over a 5,000-km. (3,105-mile) circuit at 545 m.p.h.

In the summer of 1961, three more records were set up in the height with payload category. A payload of 30,035 kg. (66,216 lb.) was carried to a height of 41,125 ft., representing a record for payloads of 20,000 kg., 25,000 kg. and 30,000 kg.

The Tu-114 is powered by four Kuznetsov NK4 turboprops believed to have a maximum output of 15,000 e.h.p. each. Contra-rotating propellers on each engine have a diameter of about 18.5 ft. Few reliable data for the Tu-114 are available. The span is 177 ft. 2 in., length 154 ft. 10 in. and height 38 ft. 8½ in.

Gross weight is in the region of 400,000 lb. and the payload about 55,000 lb. The normal cruising speed is 400 knots at 33,000 ft. but higher speeds are easily obtainable. For take-off, a runway length of about 9,400 ft. is needed.

TUPOLEV **Tu-114** (U.S.S.R.)

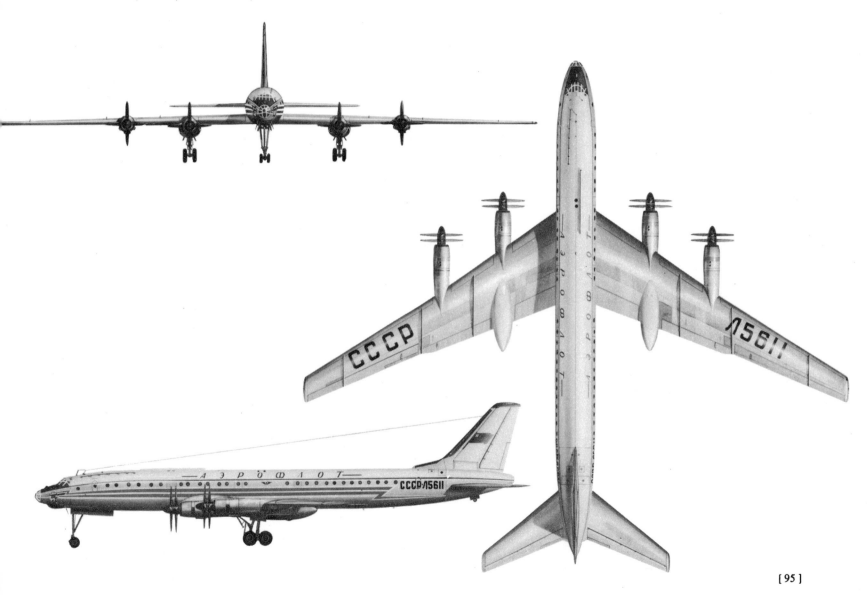

VICKERS VC10 (Great Britain)

THE BIGGEST TRANSPORT aircraft project ever undertaken in Europe, the Vickers VC10, entered the flight test stage in the summer of 1962. It is scheduled to go into service with BOAC early in 1964, about seven years after the decision to go ahead with its development.

The design history of the VC10 began in the middle of 1956 when an aeroplane known as the Vanjet was projected by Vickers for possible purchase by BEA and TCA. This design was a Vanguard fuselage with a swept-back wing and three Rolls-Royce Conways in the rear fuselage. It was not built, but influenced the Vickers designers when they projected an aeroplane to meet a BOAC requirement for a jet aircraft to operate on the Empire routes to Africa, Australia and the Far East.

During 1957, this new project, called the

---

1. *A view of the VC-10 G-ARTA on its first roll-out in April 1962.*
2. *Finished in BOAC colours, G-ARTA is a Vickers-owned demonstrator.*
3. *The VC-10 made its first flight from Weybridge in the early summer of 1962.*

VC10, progressed rapidly. The rear-engined arrangement was retained not only to obtain the required airfield and payload-range performance, but for reasons of safety, handling and maintenance.

BOAC declared its intention to order a fleet of thirty-five VC10s in May, 1957. At this time the weight was 247,000 lb. and the maximum payload was 38,000 lb. A little later the fuselage diameter was increased by 12 in. and the weight went up to 260,000 lb.

To keep the take-off distances short at tropical airfields, the VC10 was designed with an unusually low wing loading. The wing was so large, in fact, that it would carry more fuel than was needed for any of the Empire route stage-lengths. Consequently, towards the end of 1957, the specification of the basic VC10 was changed to take advantage of the extra space in the wing.

With as much fuel as could be accommodated in this space, the VC10 was found to have enough range to fly the North Atlantic. This meant that it offered BOAC the possibility of using a single type for virtually the whole of its fleet. Small changes were made in the design—to increase the wing sweepback, span and area and to substitute 18,500-lb. thrust Conway R.Co.15 engines for the R.Co.10s planned earlier.

In this guise, and at a gross weight of 299,000 lb., the VC10 was ordered by BOAC on January 14, 1958. The contract was for thirty-five with an option on twenty more and the total cost to BOAC was expected to be about £90 million.

Detailed design and production began at once, centred on the Vickers works at Weybridge. In addition to the aircraft for BOAC, the company laid down a demonstrator, and this is the aircraft (G-ARTA) rolled out in April, 1962. Early in 1959, the VC10 specification was revised to cover the use of 20,750-lb. thrust Rolls-Royce Conway R.Co.42 engines in place of R.Co.15s.

While work on the basic VC10 proceeded, larger versions were planned. As the Super VC10, a new variant was projected at weights up to 347,000 lb. with a fuselage 27 ft. longer and wing-tip tanks. On June 23, 1960, BOAC

VICKERS VC10 (Great Britain)

took up part of its earlier option with an order for ten Super VC10s of this type. Still longer and heavier versions were projected as pure freighters, with swing-nose loading facility.

A further revision in the BOAC contract, confirmed in May, 1961, reduced the number of basic VC10s on order from thirty-five to twelve (registered G-ARVA to G-ARVC and G-ARVD to G-ARVL inclusive), but increased the number of Super VC10s from ten to thirty. The size of the Super was reduced, however,

to be only 13 ft. longer than the basic type, with the gross weight pegged at 322,000 lb.

In addition to the BOAC order, Vickers have contracts from British United Airways for two VC10s (plus an option on two more) with large freight loading doors in the forward fuselage; from Ghana Airways for three and from RAF Transport Command for five. All these are 299,000-lb. basic aircraft, but a developed VC10, of "basic" size, is also available at a weight of 310,000 lb.

VC10

Dimensions
Span: 140 ft. 2 in.
Overall length: 158 ft. 8 in.
Overall height: 39 ft. 6 in.
Gross wing area: 2,800 sq. ft.
Sweepback: 32.5 deg. at quarter-chord.
Internal cabin dimensions:
   Length (ex. flight deck): 92 ft. 4 in.
   Max. width: 138 in.; max. height: 90 in.
Max. usable floor area (ex. flight deck): 1,000 sq. ft.
Max. usable volume (ex. flight deck): 6,750 cu. ft.

Accommodation
Typical mixed class: 28F (4 abreast at 40-in. pitch) and 86E (6 abreast at 40-in. pitch).
Max. high density: 150 (6 abreast at 33-in. pitch).
Volume of freight and baggage holds: 1,350 cu. ft.

Powerplants
Four Rolls-Royce Conway 540 (R.Co.42) turbofans.
Take-off power each (I.S.A., s.l.): 20,250 lb. at 9,955 r.p.m.
No water injection. Sound suppressors fitted. Thrust reversers outer engines only.

Weights and loadings
Basic operational: 138,500 lb.
Total fuel: 138,000 lb.
Mfrs. max. payload: 38,000 lb.
Max. take-off: 299,000 lb.
Max. landing: 199,000 lb.
Max. zero fuel: 178,000 lb.
Wing loading (max. t-o. wt.): 106.9 lb./sq. ft.
Wing loading (max. landing wt.): 70.5 lb./sq. ft.
Power loading (max. t-o. wt.): 3.7 lb./lb.st.

Performance
Best cost. cruising speed (I.S.A.): 480 knots at 42,000 ft. and 240,000 lb.; consumption: 1,675 imp. gal./hr.

High-speed cruise: 510 knots at 30,000 ft.
Long-range cruise: 460 knots at 42,000 ft.
Approach speed (at max. landing wt.): 122.5 knots.
Take-off field lengths (FAA requirements at max. t-o. wt.):
   At I.S.A. at sea level: 7,400 ft.
   At I.S.A. +15° C. at sea level: 8,650 ft.
   At I.S.A. at 5,000 ft.: 9,500 ft.
Landing field length (FAA requirements at max. landing wt.): 6,450 ft.
Range (take-off to landing), still air, no reserves, I.S.A., max. fuel (17,300 imp. gal.): 5,600 n. mi. with 24,500-lb. payload at 460 knots (mean) at 35,000–42,000 ft. (mean).
Range (take-off to landing), still air, no reserves, I.S.A., with max. payload: 4,800 n. mi. at 460 knots (mean) at 36,000–42,000 ft. (mean).

SUPER VC10

Dimensions
Span: 140 ft. 2 in.
Overall length: 171 ft. 8 in.
Overall height: 39 ft. 6 in.
Gross wing area: 2,800 sq. ft.
Sweepback: 32.5 deg. at quarter-chord.
Internal cabin dimensions:
   Length (ex. flight deck): 103 ft. 0 in.
   Max. width: 138 in.; max. height: 90 in.
Max. usable floor area (ex. flight deck): 1,130 sq. ft.
Max. usable volume (ex. flight deck): 7,600 cu. ft.

Accommodation
Typical mixed class: 24F (4 abreast at 42-in. pitch) and 114C (6 abreast at 33-in. pitch).
Max. high density: 163 (6 abreast at 33-in. pitch).
Volume of freight and baggage holds: 1,950 cu. ft.

Powerplants
Four Rolls-Royce Conway 550 (R.Co.42/3) turbofans.

[continued on p. 102

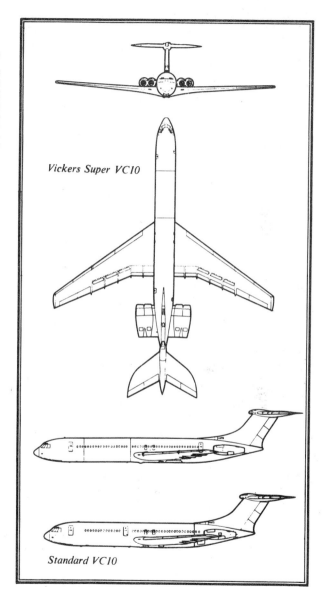

Vickers Super VC10

Standard VC10

VICKERS VANGUARD (Great Britain)

BEA Vanguard 951 G-APEB, the third Vanguard built.

VICKERS VANGUARD (Great Britain)

MORE THAN A year before the Vickers Viscount went into service with BEA, the first step had been taken towards development of its successor. As early as December, 1951, BEA planning groups had agreed that design studies should be made of such an aeroplane, known provisionally as the *Discovery* Replacement—*Discovery* being the BEA class name for the Viscount.

The kind of aeroplane required was not defined at this stage, and for the next two years a great variety of projects was studied by BEA and by Vickers, who worked closely with the airline to develop the new aeroplane. Some of the early studies were made under the designation Viscount 850, and were based on a Viscount wing with longer fuselage and uprated Darts.

These studies were followed by a series of about sixty different designs under the Vickers 870 designation, and covering high and low wings, straight and swept-back wings and turboprop and turbojet types.

When detailed design of the project began, it was designated V.900 and the name Vanguard was adopted rather than Victory, which had previously been used internally in BEA. The final BEA specification covered a payload of 21,000 lb. and accommodation for ninety-three passengers, with a gross weight of 115,000 lb.

A study of this project in relation to world markets indicated a need for better payload range performance. This was achieved by increasing the gross weight to 135,000 lb., with small structural changes in the airframe and undercarriage. Thus modified, the basic type became the V.950, and BEA concluded a contract for twenty V.951s on July 20, 1956.

Further strengthening of the airframe made it possible to increase the payload to 24,000 lb. without affecting the range. The gross weight went up to 141,000 lb. and landing and zero fuel weights increased proportionally. This version, as the V.952, was ordered in January, 1957, by TCA, which later increased the quantity on order from twenty to twenty-three. By adopting a six-abreast layout, the accommodation in this version was increased to 120.

Whereas the BEA Vanguards were specified with Tyne R.Ty.1s, which had adequate power for the lower gross weight, TCA ordered R.Ty.11s.

A further increase in payload was announced in July, 1958, when the Vanguard II was offered with a payload of 29,000 lb. for ranges up to 2,000 st. miles. No change was made in gross weight, which was 141,000 lb., but landing and zero fuel weights again increased. BEA decided to take advantage of this development in the final fourteen of its fleet, which became V.953s. In 1960, a year after the Vanguard's first flight, the gross weight of the V.952 and V.953 was increased to 146,500 lb. and the payloads were further improved, to 33,500 lb. and 37,000 lb. respectively.

Vickers put the Vanguard into production at Weybridge and built a demonstrator in addition to the forty-three ordered by BEA and TCA. This demonstrator, G-AOYW, made its first flight on January 20, 1959, quickly followed by the first of the BEA V.951s. An early modification was to increase the size of the dorsal fin, but otherwise few changes were made and test flying continued smoothly until June, 1960, by which time eight aircraft had flown 1,400 hours and almost all the work for the C. of A. was complete.

A serious fault in the compressors of certain Tyne engines was then discovered, and further flying with this engine was stopped until the fault could be rectified. This kept the Vanguards on the ground from May 23 to July 4. When flights resumed, a 250-hour engine-assurance flight programme had to be completed and the Vanguard was not certificated for unrestricted passenger carrying until December 2, 1960.

On December 17, the first Vanguard service was flown by BEA, using G-APEE which had been delivered on December 3. Services from then until March were on an *ad hoc* basis, however, full service on the Paris route starting

---

1. *The Vanguard prototype G-AOYW as first flown.*
2. *G-AOYW with dorsal fin added.*
3. *BEA'S first Vanguard 951 in original colours and without the dorsal fin.*
4. *The Viscount 812 G-AOYY with a Vanguard fin for de-icing trials.*

VICKERS VANGUARD (Great Britain)

VICKERS VANGUARD (Great Britain)

on March 1, 1961. The last BEA Vanguard, a V.953, was delivered on March 30, 1962.

TCA, meanwhile, had taken delivery of their first V.952 at the end of 1960 and inaugurated regular scheduled service on February 1, 1961.

5. *The second Vanguard 951 for BEA, G-APEB.*

6. *TCA's second Vanguard 952, CF-TKB.*

7. *The first TCA Vanguard 952, CF-TKA.*

VICKERS VANGUARD

Dimensions

Span: 118 ft. 0 in.
Overall length: 122 ft. 10½ in.
Overall height: 34 ft. 11 in.
Gross wing area: 1,529 sq. ft.
Sweepback: Nil.
Internal cabin dimensions:
  Length (ex. flight deck): 90 ft. 0 in.
  Max. width: 129 in.; max. height: 82.5 in.
Max. usable floor area (ex. flight deck): 893 sq. ft.
Max. usable volume (ex. flight deck): 5,690 cu. ft.

Accommodation

Typical mixed class: 42F (4 abreast at 38-in. pitch) and 55E (6 abreast at 38-in. pitch).
Max. high density: 139 (6 abreast at 33-in. pitch).
Volume of freight and baggage holds: 1,360 cu. ft.

Powerplants

Four Rolls-Royce Tyne 512 (R.Ty.11) turbo-props.
Take-off power each (I.S.A., s.l.): 5,545 e.h.p. at 15,250 r.p.m.
No water injection. De Havilland four-blade 14-ft. 6-in. reversing propellers.

Weights and loadings

Basic operational: 85,500 lb.
Total fuel: 41,040 lb.
Mfrs. max. payload: 37,000 lb.
Max. take-off: 146,500 lb.
Max. landing: 130,500 lb.
Max. zero fuel: 122,500 lb.
Wing loading (max. t-o. wt.): 96.0 lb./sq. ft.
Wing loading (max. landing wt.): 88.3 lb./sq. ft.
Power loading (max. t-o. wt.): 6.6 lb./e.h.p.

Performance

Best cost. cruising speed (I.S.A.): 358 knots at 20,000 ft. and 135,000 lb.; consumption: 710 imp. gal./hr.
High-speed cruise: 367 knots at 15,000 ft.
Long-range cruise: 356 knots at 25,000 ft.
Approach speed (1.3V_{s0} at max. landing wt.): 131.5 knots.

Take-off field lengths (FAA requirements at max. t-o. wt.):
  At I.S.A. at sea level: 6,550 ft.
  At I.S.A. +15° C. at sea level: 7,950 ft.
  At I.S.A. at 5,000 ft.: 9,000 ft.
Landing field length (FAA requirements at max. landing wt.): 6,400 ft.
Range (take-off to landing), still air, no reserves, I.S.A., max. fuel (5,130 imp. gal.): 2,720 n. mi. with 20,500-lb. payload at 356 knots (mean) at 25,000 ft. (mean).
Range (take-off to landing), still air, no reserves, I.S.A., with max. payload: 1,590 n. mi. at 350 knots (mean) at 25,000 ft. (mean).

Structure. Double-bubble monocoque fuselage. Torsion-box wing with three shear webs and integrally machined upper and lower skin-stringer panels.

Fuel system. Two integral tanks in each wing. Capacity, 5,100 imp. gal. Single-point pressure refuelling at 600 g.p.m.

Undercarriage. Forwards retracting main and steerable nose oleos. Track, 30 ft. 3 in. Base, 40 ft. 6¼ in. Tyre pressure, 105–123 p.s.i.

Flying controls. Manually operated through push-pull rods with aerodynamic balance and spring tabs. Collins AP-103 or Smiths S.E.P.2 auto-pilot.

Flaps, air brakes. Constant-chord Fowler-type flaps in four sections per side. No air brakes.

Cabin conditioning. Two engine-driven blowers. Cooling by air-cycle refrigeration. Differential, 6.5 p.s.i.

Hydraulics. Duplicated 3,000-p.s.i. system powered by four engine-driven pumps for u/c, flaps, brakes, nosewheel steering, airstairs and propeller brakes.

Electrics. Six variable frequency 50-kVA. alternators, provide 208-V. a.c. and, via four transformer/rectifiers, 28-V. d.c. for main electrical demand.

De-icing. Wings by engine heat exchangers. Alternators on Nos. 1 and 4 engines provide current for tail unit, Spraymats and Triplex windscreen. Electric cycle for propellers, spinners and engines.

5

6

7

SUPER VC10 *(continued from page 97)*

Take-off power each (I.S.A., s.l.): 22,500 lb. at 10,172 r.p.m.

No water injection. Sound suppressors fitted. Thrust reversers outer engines only.

Weights and loadings
Basic operational: 152,500 lb.
Total fuel: 152,500 lb.
Mfrs. max. payload: 46,500 lb.
Max. take-off: 322,000 lb.
Max. landing: 225,000 lb.
Max. zero fuel: 200,000 lb.
Wing loading (max. t-o. wt.): 115.0 lb./sq. ft.
Power loading (max. t-o. wt.): 3.58 lb./lb. st.

Performance
Best cost. cruising speed (I.S.A.): 490 knots at 36,000 ft. and 230,000 lb.
High-speed cruise: Mach 0.86 at 30,000 ft.
Long-range cruise: Mach 0.815 knots at 42,000 ft.
Approach speed ($1.3 V_{s_0}$ at max. landing wt.): 130 knots.
Take-off field lengths (BCAR requirements at max. t-o. wt.):
  At I.S.A. at sea level: 8,200 ft.
  At I.S.A. +15° C. at sea level: 9,500 ft.
  At I.S.A. at 5,000 ft.: 10,600 ft.
Landing field length (BCAR requirements at max. landing wt.): 7,300 ft.
Range (take-off to landing), still air, no reserves, I.S.A., max. fuel (19,065 imp. gal.): 5,600 n. mi. with 17,050-lb. payload at 464 knots (mean) at 40,000 ft. (mean).
Range (take-off to landing), still air, no reserves, I.S.A., with max. payload: 4,235 n. mi.

Structure. Double-bubble monocoque fuselage. Torsion-box wing in five sections, with centre section built into lower half of fuselage. Four shear webs and integral-machined wing panels.

Fuel system. Three integral tanks in each wing and one in centre section, between spars. Capacity, 17,300 imp. gal. Two-point pressure refuelling at 1,000 g.p.m.

Undercarriage. Four-wheel main bogies, inwards retracting; twin-wheel steerable nosewheel, forwards retracting. Track, 21 ft. 3 in. Base, 66 ft. Tyre pressure, 115 p.s.i.

Flying controls. Fully powered, with sectionalized surfaces and individual actuators for each section operated by duplicated independent control systems. Artificial feel. Elliott-Bendix PB20 autopilot.

Flaps, air brakes. In five sections per side. Full span leading edge slats. Three spoilers above each wing.

Cabin conditioning. Four Lysholm engine-driven compressors. Freon vapour-cycle conditioning and humidification system. Liquid oxygen emergency system. Differential, 9.0 p.s.i.

Hydraulics. Two independent 3,000-p.s.i. systems for u/c, brakes, nosewheel steering, slats, spoilers and variable-incidence tailplane.

Electrics. Four 40-kVA. constant-speed engine-driven generators to give 115/200-V. 400-cycle 3-phase a.c.

De-icing. Wing, tail and powerplant by engine-bleed air. Nesa-glass windscreen.

OF THE 2,000 or so turbine-engined airliners purchased by the world's airlines up to the middle of 1962, more than 400 are Vickers Viscounts. First flown in July, 1948, the Viscount has been built in greater numbers than any other airliner with turbojet or turboprop engines and it is one of the very small number of post-war civil aircraft which have proved profitable for their makers.

Design work which led eventually to production of the Viscount began before the end of 1944. The Vickers design staff (led by the late Rex K. Pierson) had at that time completed preliminary designs of a transport version of the Wellington known as the VC1 Viking. Discussion in the Brabazon Committee and at Vickers led to the conclusion that a twenty-four-seat turboprop transport should be developed to succeed the Viking on short-haul routes in Europe.

This type became known as the Brabazon IIB requirement and by March, 1945, future proposals had been made to the Ministry of Supply for work on such an aeroplane to begin. Vickers' project designs at this stage were made under the drawing number 453 and three possibilities for a VC2 transport were proposed in March, 1945. These were for pressurized aeroplanes weighing 35,000 lb. and with seats for twenty-four to twenty-seven passengers.

The Brabazon IIB requirements were set out in greater detail in May, and asked for twenty-four seats in a pressurized fuselage, four engines, and a 900 n. mi. range with a 7,500-lb. payload. A variant of Vickers Type 453 to meet these requirements had a "double-bubble" fuselage cross section, three-abreast

seating and the elliptical doors and windows which remained right through the project stages until the Viscount was built.

Various engines were considered during 1945 but before the end of that year George (now Sir George) Edwards, who had succeeded Pierson as chief designer, decided to use the Rolls-Royce Dart. Most of the basic design decisions were also made during 1945. Compared with the Type 453 projects of mid-year, the VC2 ended the year as an aeroplane weighing 35,500 lb. with a circular-section fuselage, main undercarriage retracting forwards instead of inwards, a dihedral tailplane and equal taper on the wing leading and trailing edges. The span was 55 ft. and the length 65 ft. 5 in. The required 7,500-lb. payload could be carried for 900 n. mi. at 256 knots at 20,000 ft.

On March 9, 1946, Vickers received a contract from the Ministry of Supply to build prototypes of this design to Specification 8/48. Accommodation was to be increased to thirty-two, and the span and length were increased, in consequence, to 89 ft. and 74.5 ft. For some time there was uncertainty about the engines, with the Armstrong Siddeley Mamba competing with the Dart, but Vickers' own choice of the latter was eventually confirmed.

Originally the Ministry was to order four prototypes but this was reduced later to two, with Vickers financing a third. On Ministry instruction, the first two aircraft were to have Mamba engines and took the Type number 609. The name Viceroy, adopted after construction began, was changed after the partition of India made it inappropriate.

During 1946, versions were studied with a variety of different powerplants, including two Napier Double Naiads and four Naiads, but progress with the Dart eventually led the Ministry to specify these engines for the second prototype, which then became V630; the first prototype subsequently was converted, before the first flight, to the same standard. The Vickers-financed third prototype was to be a V640 with Naiad engines.

British European Airways, as potentially the obvious first customer for the Viscount, had been closely linked with its development from mid-1945, but a production order was not placed for more than five years. For a time,

VICKERS **VISCOUNT 700** (Great Britain)

N6592C, the first Viscount 798 operated by Northeast Airlines.

the Viscount appeared commercially unattractive and schemes were even studied for a version with piston engines. The first flight of the first prototype (G-AHRF) on July 16, 1948, did little to help and Vickers abandoned construction of their own prototype whilst also slowing down work on the second.

During 1948, project design work on the Viscount led to a proposal for an enlarged version, taking advantage of development of the Dart to have 50 per cent. more power than in the V630. With this additional power, Vickers were able to offer a Viscount with a larger fuselage and seats for forty passengers. Economically, this was much more attractive to BEA than the V630, and the airline pre-

vailed upon the Ministry to order a prototype of the larger version.

By placing a new contract for the second V630 to be completed as a flying test bed for the Rolls-Royce Tay turbojet, funds were freed for this new prototype which was studied as the V655 and then became the V700. This was built to Specification 21/49 and its new features were a 60-in. increase in span, 80-in. longer fuselage, revised nacelles with "petal" cowlings, a new fuel system and a weight of 50,000 lb.

Registered G-AMAV, this prototype was first flown on April 19, 1950, only eighteen months after it had been ordered. While test flights with the V700 were in progress, the

1. *Viscount 630, the first of the type, with RAF markings and serial number of VX211.*
2. *The Viscount 700 prototype, G-AMAV.*
3. *First production-type Viscount, the V701 G-AMOG for BEA.*
4. *Viscount 702 VP-TBU for BWIA.*
5. *An Aer Lingus Viscount 708 EI-AGI.*
6. *An ex-BWIA Viscount 702, G-APPX, for Bahamas Airways.*
7. *Air France Viscount 707 F-BGNK.*
8. *Carrying long-range tanks, the Viscount 720 VH-TVH for TAA.*
9. *CF-TGI, the first Viscount 724 for TCA.*
10. *In Middle East Airlines' livery, the Viscount 732 OD-ACW.*
11. *Iraqi Airways' Viscount 735 YI-ACK.*
12. *Viscount 739 SU-AIC in Egyptian Airlines markings.*
13. *A late Viscount 745 for Capital Airlines, N7462.*
14. *United Air Lines' Viscount 745 N7408.*
15. *CAA's Viscount 748 VP-YNE with slipper tanks.*
16. *The Fred Olsen Viscount 779 on charter to SAS.*

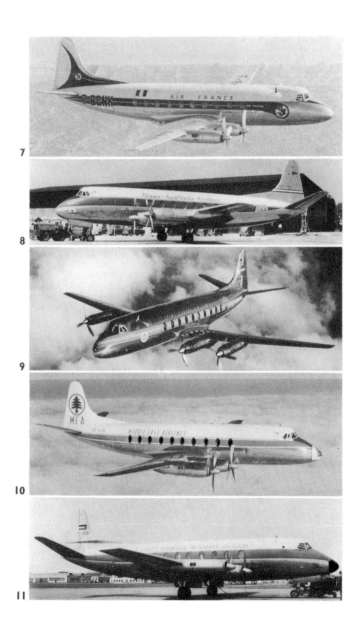

7

8

9

10

11

12

13

14

15

16

17

18

19

20

21

V630 was used—from July 29, 1950, for a month—for scheduled passenger service with BEA on routes from London to Edinburgh and Paris. These were the first turbine-engined flights for fare-paying passengers in the world.

The BEA order for twenty Viscounts was confirmed on August 3, 1950, and was for aircraft known as V701s. These were in most respects similar to the V700 prototype but with accommodation for fifty-three passengers. They were powered by Dart R.Da.3s at Mk. 505 rating and had an initial design weight of 56,000 lb. This later increased to 60,000 lb. after the engine rating had been increased. New features of the production aircraft were seat rails to obtain maximum flexibility of cabin layouts, and under-floor freight holds.

BEA put the Viscount 701 into service on April 17, 1953. Excluding the month-long evaluation of the V.630, this marked the beginning of regular turboprop transport operations for passengers; turbojet operation by the de Havilland Comet had already begun.

Once BEA had taken the initiative, orders for the Viscount began to accumulate, and so did the variants. Under Vickers' designating system, different design numbers were allocated to every separate airline order; in some cases, a single airline ordered more than one version, and many other numbers covered projects and tentative orders which did not materialize. A summary of airline and other orders appears below, but second-hand deals over the past few years have complicated the

situation, as a close study of the Production Record will show.

In many cases, the different designations represent little more than a difference in equipment or furnishing standard. Certain variants, however, are of more consequence, including the larger 800 Series described separately on pp. 110–113. All variants with numbers between 700 and 799 have the same size of airframe.

To meet Trans-Australia Airlines' requirements, the fuel capacity of the 700 Series Viscounts was increased in the V720, by introducing extra tanks in the inner wing and provision for external "slipper" tanks outboard of the outer nacelles. The extra fuel was specified by several later customers.

Another series of design changes—which actually required more man-hours than did the work on the original V630—was made for Trans-Canada Air Lines, which was the first airline in North America to order the Viscount. The changes in this case included a great deal of American equipment and styling, a new fuel system, a cockpit suitable for two-crew operation and a revised undercarriage for a gross weight of 60,000 lb. This version was the V724.

In 1954, Vickers obtained an order for Viscounts from Capital Airlines. To obtain certification in the U.S., further changes had to be made, but much of the work had already been done for TCA. Special features for Capital in the V745 included a Freon cabin

---

17. The LAV Viscount 749, YV-C-AMX.
18. Cubana's Viscount 755 CU-T604 before sale to Eagle.
19. TF-ISU, Icelandair Viscount 759.
20. The Hong Kong Airways Viscount 760 VR-HFJ with slipper tanks.
21. Indian Airlines' Viscount 768 VT-DJC.
22. An ex-Air France Viscount 708, F-BGNU, in Air Viet Nam markings.
23. Another Viscount 708, ex-Air France, supplied to Starways as G-ARIR.
24. The second Viscount 747, VH-BUT, in original Butler markings.

25. The first Butler Viscount 747, VH-BAT, repainted in Ansett-ANA colours.
26. EP-AHA, a Viscount 782, in the livery of United Iranian Airlines.
27. An ex-Airwork, ex-Cubana, Viscount 755 as G-ARKH in Cunard Eagle markings.
28. Maitland Drewery's Viscount 708, G-ARBY, later leased to Silver City.
29. The Viscount 708, G-ARER, leased by BKS from Maitland Drewery.
30. Ex-BWIA Viscount 702 as G-APTA for Kuwait Airways.
31. Ex-MEA Viscount 754 as JY-ACI for Jordan Airways.

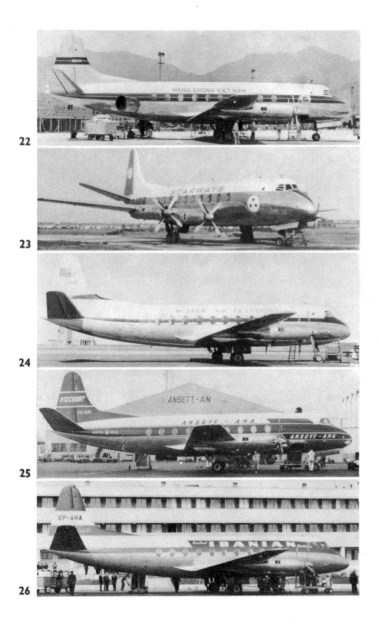

22

23

24

25

26

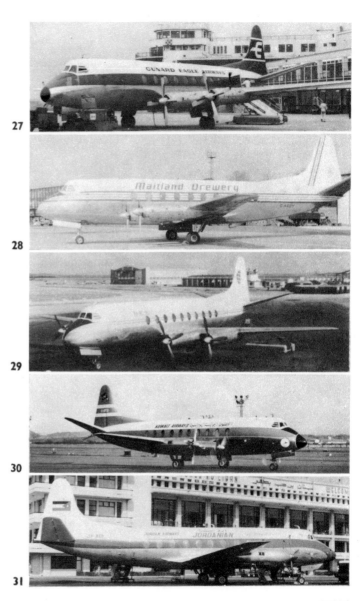

27

28

29

30

31

32

33

34

35

36

conditioning system, weather radar, integral air-stairs, fuel jettisoning and a new wing spar to give adequate strength and life on low-altitude, short-range operation.

The V745 was also the first Viscount to make use of the Dart R.Da.6 series of engines, offering 200 e.h.p. more each. With these engines, the gross weight increased to 63,000 lb. and eventually to 64,500 lb. First FAA certification of the Viscount was obtained on June 13, 1955. This was for the V744, an interim model for Capital. The V745 was FAA-approved on November 7, 1955, and the V745D on February 23, 1956.

Combining features of the V720, V724 and V745 in a new standard type, Vickers offered the V700D Series to later customers. This designation applied to any Viscount 700s with

R.Da.6 engines and standard airframe components. Another designation—V770D—was reserved for basic North American versions of the V700D, while the V771D applied to executive versions of the V770D. Yet another Series designation was V790 for the version offered to local-service airlines in the U.S., with special features for quick turn-rounds and short-stage flights, and accommodation for up to sixty-five passengers.

The list of "customer" type numbers between 700 and 799 follows.

V701. Twenty-seven for BEA with Dart 506.

V702. Four for BOAC Associated Companies. All allocated to BWIA; three later to Kuwait Airways and two of these later to Bahamas Airways.

V707. Four for Aer Lingus. Two later to Tradair, two to Eagle (later Cunard Eagle) Bermuda and Bahamas companies, then to Cunard Eagle parent company.

V708. Twelve for Air France. Two later to Air Viet-Nam; five to Air Inter; one to Starways; three to Maitland Drewery. Two of Maitland Drewery leased by BKS and one by Danish Air Charter; all three later leased by Silver City.

V720. Seven for TAA. Three later leased to Ansett-ANA.

V723. One Indian Air Force VIP.

V724. Fifteen for TCA. See V757 also.

V730. One Indian Air Force VIP.

---

32. *The PLUNA Viscount 769 CX-AQN.*
33. *Philippine Air Lines' Viscount 784, PI-C770.*
34. *The Viscount 785 I-LIZT repainted in Alitalia markings.*
35. *Lloyd Aereo Colombiano's Viscount 786 HK-943X before resale.*
36. *The Lloyd Viscount 786 in LANICA markings as AN-AKQ.*
37. *Viscount 760 in Malayan Airways colours as 9M-ALY.*
38. *Turkish Airlines' Viscount 794 TC-SEC.*

37

VICKERS VISCOUNT 700 (Great Britain)

V732. Three for Hunting Clan. Leased to MEA. One later leased to BEA; one leased to Iraqi Airways. Two later sold to Misrair (UAA).

V734. One Pakistan Government VIP.

V735. Three for Iraqi Airways.

V736. Two for Fred Olsen. Leased to BEA. One later leased to MEA and the other to BWIA. Both sold to Airwork (BUA).

V737. One for Canadian Department of Transport.

V739. Six for Misrair.

V742. One ordered by Braathens, sold to Brazilian Government.

V744. Three for Capital Airlines on lease. Two returned to Britain, leased to All Nipon, and one then sold to Empire Test Pilot's School.

V745. Sixty for Capital. Forty-seven to United Air Lines after the merger, five to Alitalia, two to Austrian Airlines, one to Philippine Air Lines, one to Empire Test Pilots' School.

V747. Two for Butler Air Transport, later repainted in colours of the parent Ansett-ANA company.

V748. Five for Central African Airways. One leased to Kuwait.

V749. Three for LAV.

V754. Seven for BOAC Associated Companies, allocated to MEA. Two later leased to Jordanian, one sold to Central African.

V755. Three ordered by Airwork, sold to Cubana. Two later sold to Cunard Eagle Bermuda and Bahamas companies.

V756. Eight for TAA.

V757. Thirty-six for TCA. See V724 also.

V759. Two ordered by Hunting Clan, sold to Icelandair.

V760. Two for BOAC Associated Companies allocated to Hong Kong Airways. Later transferred to Malayan Airway.

V761. Three for Union of Burma Airways. One leased to Kuwait Airways.

[continued overleaf

38

VICKERS VISCOUNT 700D

Dimensions
Span: 93 ft. 8½ in.
Overall length: 81 ft. 10 in.
Overall height: 36 ft. 9 in.
Gross wing area: 963 sq. ft.
Sweepback: Nil.
Internal cabin dimensions:
  Length (ex. flight deck): 45 ft. 0 in.
  Max. width: 120 in.; max. height: 78 in.
Max. usable floor area (ex. flight deck): 430 sq. ft.
Max. usable volume (ex. flight deck): 2,370 cu. ft.

Accommodation
Typical first class: 44 (4 abreast).
Max. high density: 63 (5 abreast).
Volume of freight and baggage holds: 215 cu. ft.

Powerplants
Four Rolls-Royce Dart 510 (R.Da.6) turboprops.
Take-off power each (I.S.A., s.l.): 1,740 e.h.p.
Water methanol system used. Rotol four-blade 10-ft. diameter reversing propellers.

Weights and loadings
Basic operational: 38,358 lb.
Total fuel: 15,736 lb.
Mfrs. max. payload: 11,600 lb.
Max. take-off: 64,500 lb.
Max. landing: 57,500 lb.

Max. zero fuel: 50,168 lb.
Wing loading (max. t-o. wt.): 67.0 lb./sq. ft.
Wing loading (max. landing wt.): 59.7 lb./sq. ft.
Power loading (max. t-o. wt.): 9.26 lb./e.h.p.

Performance
Best cost. cruising speed (I.S.A.): 270 knots at 20,000 ft. and 57,500 lb.; consumption: 335 imp. gal./hr.
High-speed cruise: 276 knots at 12,000 ft.
Long-range cruise: 262 knots at 25,000 ft.
Approach speed ($1.3 V_{s_0}$ at max. landing wt.): 115 knots.
Take-off field lengths (FAA requirements at max. t-o. wt.):
  At I.S.A. at sea level: 5,280 ft.
  At I.S.A. +15° C. at sea level: 5,760 ft.
  At I.S.A. at 5,000 ft.: 7,120 ft.
Landing field length (FAA requirements at max. landing wt.): 2,400 ft.
Range (take-off to landing), still air, no reserves, I.S.A., max. fuel (1,967 imp. gal.): 1,740 n. mi. with 10,680-lb. payload at 269 knots (mean) at 21,000 ft. (mean).
Range (take-off to landing), still air, no reserves, I.S.A., with max. payload: 1,550 n. mi. at 269 knots (mean) at 21,000 ft. (mean).
Engineering Summary—as for Viscount 810 on p. 113.

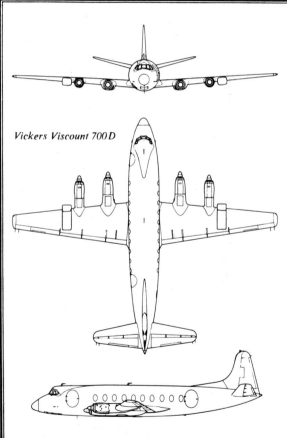

Vickers Viscount 700 D

VICKERS VISCOUNT 800 (Great Britain)

continued from p. 109]

V763. One ordered by Howard Hughes, sold to TACA.

V764. Three executive for United States Steel Corp.

V765. One executive for Standard Oil Co.

V768. Ten for Indian Airlines Corp.

V769. Three for PLUNA.

V772. Four for BOAC Associated Companies allocated to BWIA.

V773. One for Iraqi Airways.

V776. One for Gulf Aviation.

V779. Four for Fred Olsen, leased to Austrian Airlines. Two leased to BEA, two leased to SAS. All later sold to Indian Air lines.

V781. One VIP for South African Government.

V782. Three for Iranian Airlines.

V784. Two for Philippine Airlines. One leased to TACA.

V785. Ten for Alitalia.

V786. Three ordered by Lloyd Aereo Columbiano; one sold to TACA and two sold to LANICA, of which one later sold to TACA and one to Mrs. H. M. May.

V793. One for Royal Bank of Canada. Leased before delivery to Aer Lingus, BEA and Kuwait Airways. Later sold to Canadian Breweries, then S. J. Groves and Son.

V794. Five for BOAC Associated Companies allocated to THY.

V797. One for Canadian Department of Transport.

V798. Ten for Northeast Airlines.

IN 1952, VICKERS proposed an enlarged version of the Viscount, designed around the up-rated 1,690-e.h.p. Dart R.Da.5. The "stretch" in this version was primarily in payload. By lengthening the fuselage by 13 ft. 3 in. it was possible to provide for as many as eighty-six passengers. At a gross weight of 65,000 lb., this version—called the V800 Series—had a cruising speed a little below 300 m.p.h., whereas the V700D Series cruised at 326 m.p.h. and the original V700 figure was 300 m.p.h.

On February 11, 1953, BEA placed an order for twelve of the enlarged Viscounts, which took the customer designation V801. Further consideration of the project, however, both by Vickers and by BEA, led to the conclusion that the "stretch" was too great.

In the course of the next year, the V800 Series was redesigned, therefore, around the still more powerful 1,740-e.h.p. Dart R.Da.6, which was already being used in the V700D. The fuselage lengthening was reconsidered and a figure of 3 ft. 10 in. was eventually accepted, although the effective cabin length was increased by 9 ft. 3 in. by relocating the rear pressure bulkhead further aft.

Thus redesigned, the new Viscount seated sixty-five passengers and with a gross weight of 63,000 lb. it cruised at 320 m.p.h. BEA changed its order to the new variant on April 14, 1954, taking up an option on ten more a year later. The designation for BEA changed to V802 and apart from the longer fuselage, a distinguishing feature was the use of rectangular rather than elliptical main cabin doors.

---

1. *KLM Viscount 803 PH-VIA, the first Series 800 for export.*
2. *Transair's Viscount 804 G-AOXU, before joining the BUA fleet.*
3. *Eagle Airways' Viscount 805 G-APDW subsequently joined the Aer Lingus fleet.*
4. *A BEA Viscount 806, G-AOHL.*
5. *The NZNAC Viscount 807, ZK-BRD.*
6. *Viscount 808 EI-AJJ for Aer Lingus.*
7. *Continental Airlines' first Viscount 812, N240V—the first of fifteen ordered.*
8. *The ex-Airwork Viscount 831 G-APNE in BUA colours.*
9. *BUA's Viscount 833 G-APTC on lease to BOAC.*
10. *The VASP Viscount 827 PP-SRC.*
11. *Austrian Airlines' Viscount 837 OE-LAF.*
12. *The Hunting Clan Viscount 833 G-APTB before being repainted in BUA livery.*
13. *TAA's Viscount 816 VH-TVQ.*

6

7

8

9

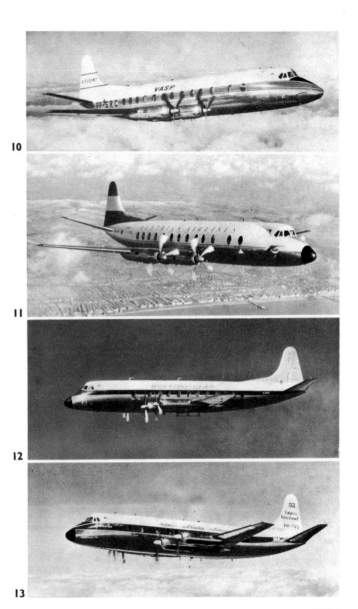

10

11

12

13

VICKERS VISCOUNT 800 (Great Britain)

[112]

VICKERS VISCOUNT 800 (Great Britain)

The V800 Series Viscount was further developed into the V810, which had a strengthened airframe for 2,100-e.h.p. Dart R.Da.7 (Mk. 525) engines, gross weight increased to 69,000 lb. and higher cruising speeds. After going into service, the V810 was cleared for operation at weights up to 72,000 lb., or more than double the weight of the original VC.2 project. FAA approval for operations in the U.S. was obtained on April 22, 1958.

[continued on p. 130

VICKERS VISCOUNT 810

Dimensions
Span: 93 ft. $8\frac{1}{2}$ in.
Overall length: 85 ft. 8 in.
Overall height: 26 ft. 9 in.
Gross wing area: 963 sq. ft.
Sweepback: Nil.
Internal cabin dimensions:
   Length (ex. flight deck): 54 ft. 0 in.
   Max. width: 120 in.; max. height: 78 in.
Max. usable floor area (ex. flight deck): 520 sq. ft.
Max. usable volume (ex. flight deck): 2,800 cu. ft.

Accommodation
Typical first class: 52 (4 abreast at 39-in. pitch).
Max. high density: 70 (5 abreast).
Volume of freight and baggage holds: 250 cu. ft.

Powerplants
Four Rolls-Royce Dart 525 (R.Da.7/1) turbo-props.
Take-off power each (I.S.A., s.l.): 1,990 e.h.p. at 15,000 r.p.m.
No water injection. Rotol four-blade 10-ft. diameter reversing propeller.

Weights and loadings
Basic operational: 43,000 lb.
Total fuel: 15,200 lb.
Mfrs. max. payload: 14,500 lb.
Max. take-off: 72,500 lb.
Max. landing: 64,000 lb.
Max. zero fuel: 57,500 lb.
Wing loading (max. t.-o. wt.): 75.4 lb./sq. ft.
Wing loading (max. landing wt.): 66.4 lb./sq. ft.
Power loading (max. t.-o. wt.): 9.11 lb./e.h.p.

Performance
Best cost. cruising speed (I.S.A.): 309 knots at 18,000 ft. and 60,000 lb.; consumption: 400 imp. gal./hr.
High-speed cruise: 317 knots at 17,000 ft.
Long-range cruise: 298 knots at 25,000 ft.
Approach speed ($1.3 V_{s_0}$ at max. landing wt.): 121.5 knots.
Take-off field lengths (FAA requirements at max. t.-o. wt.):
   At I.S.A. at sea level: 5,930 ft.
   At I.S.A. +15° C. at sea level: 6,480 ft.
   At I.S.A. at 5,000 ft.: 7,950 ft.

Landing field length (FAA requirements at max. landing wt.): 4,450 ft.
Range (take-off to landing), still air, no reserves, I.S.A., max. fuel (1,900 imp. gal.): 1,530 n. mi. with 14,300-lb. payload at 290 knots (mean) at 24,000 ft. (mean).
Range (take-off to landing), still air, no reserves, I.S.A., with max. payload: 1,500 n. mi. at 290 knots (mean) at 24,000 ft. (mean).

Structure. Semi-monocoque circular-section fuselage. Single-spar wing with two subsidiary spanwise members to complete torsion box, built in five sections.

Fuel system. Two bag tank groups in each wing. Capacity, 1,900 imp. gal. Optional 145-imp. gal. slipper tanks. Two-point pressure refuelling at 400 g.p.m.

Undercarriage. Twin-wheel main and steerable nose oleos. Track, 25 ft. 5 in. Base, 28 ft. $8\frac{1}{2}$ in. Tyre pressure, 120 p.s.i.

Flying controls. Manually operated by push-pull rods. Smiths S.E.P. autopilot.

Flaps, air brakes. Double-slotted forward-retracting type in six sections, electrically operated. No air brakes.

Cabin conditioning. Three engine-driven blowers. Cooling by turbo-expander and air-cycle refrigeration. Differential, 6.5 p.s.i.

Hydraulics. Duplicated hydraulic system at 2,000–2,500 p.s.i., powered by two engine-driven pumps, for u/c, nosewheel steer and brakes.

Electrics. Four 9-kW. generators.

De-icing. De-icing of wing and tail surfaces by hot air from engine heat exchangers; electric propeller and intake de-icing.

---

14. *South African Airways' Viscount 813 ZS-CDT.*
15. *A Pakistan International Viscount 815, AP-AJC.*
16. *Ghana Airways Viscount 828 9G-AAV.*
17. *A Lufthansa Viscount 814, D-ANUN.*
18. *The Ansett-ANA Viscount 832, VH-RMH.*

14

15

16

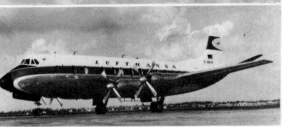

17

18

PRODUCTION RECORD

AIRCRAFT ARE LISTED in numerical sequence of their constructor's number (**bold type**); this is not necessarily the order in which the aircraft were built. Registrations in the second column (also in bold type) are those currently carried; earlier markings are mentioned in the Comments column. Individual aircraft names are shown in italics; but without common fleet names, which are as follows:

CPA—all names have *Empress* prefix.

PAA—all names have *Clipper* prefix.

American Airlines—all names have *Flagship* prefix. Names have been deleted from the Boeing 707/720 fleet on conversion to turbofans; those originally allocated are shown.

SAS—all names have *Viking* suffix.

Air France—Boeing 707 names have *Château de* prefix.

BEA—Vanguard and Comet names are carried only inside the aircraft. Viscount names have been deleted when the aircraft are repainted in the current colour scheme.

Abbreviations: Proto.—prototype; Srs.—series; F.F.—first flight; Cvtd.—converted; D.—delivered; L.—leased; S.—sold; Sc.—scrapped; W.O.—written off.

ARMSTRONG-WHITWORTH ARGOSY

6651	**G-AOZZ**	F.F. 8.1.59. Cvtd. as Srs.102 for BEA and D. 19.12.61.
6652	**N6507R**	F.F. as G-APRL 14.3.59. D. Riddle 17.8.61.
6653	**G-APRM**	F.F. 26.4.59. Cvtd. as Srs.102 for BEA and D. 23.11.61.
6654	**G-APRN**	F.F. 13.5.59. Cvtd. as Srs.102 for BEA F.F. 21.9.61 and D. 1.11.61.
6655	**N6504R**	F.F. as G-APVH 20.7.59. D. Riddle 22.6.61.
6656	**N6503R**	F.F. as G-APWW (ex G-1-3) 23.9.59. D. Riddle 24.1.61.
6657	**N6505R**	F.F. 25.2.61 (ex G-1-4). D. Riddle 11.6.61.
6658	**N6506R**	F.F. 26.6.61 (ex G-1-5). D. Riddle 18.7.61.
6659	**N6501R**	F.F. 18.11.60 (ex G-1-6). D. Riddle 12.12.60.
6660	**N6502R**	F.F. 16.12.60 (ex G-1-7). D. Riddle 7.1.61.

AVRO 748

1534	**G-APZV**	1st Proto.: F.F. 24.6.60.
1535	**G-ARAY**	2nd Proto.: F.F. 10.4.61. Cvtd. Srs.2. F.F. 6.11.61. Pride of Perth.
1536	**G-ARMV**	Srs.1: F.F. 30.8.61. D. Skyways 3.62.
1537	**G-ARMW**	Srs.1: D. Skyways.
1538	**G-ARMX**	Srs.1: D. Skyways.
1539	**LV-PIZ**	Srs.1: F.F. 10.12.61. D. Aerolineas 11.1.62. Ciudad de Bahia Blanca.
1540	**LV-PJA**	Srs.1: F.F. 22.2.62. D. Aerolineas.
1541	**LV-PJR**	Srs.1: D. Aerolineas.
1542	**LV-PJS**	Srs.1: D. Aerolineas.
1543	**LV**	Srs.1: D. Aerolineas.
1544	**LV**	Srs.1: D. Aerolineas.
1545	**LV**	Srs.1: D. Aerolineas.
1546	**LV**	Srs.1: D. Aerolineas.
1547	**LV**	Srs.1: D. Aerolineas.
1548	**G-ARRV**	Srs.2: D. Brazilian Air Force.
1549	**G-ARRW**	Srs.2: D. Brazilian Air Force.
1550	**VR-AAU**	Srs.2: D. Aden Airways.
1551	**VR-AAV**	Srs.2: D. Aden Airways.
1552	**VR-AAW**	Srs.2: D. Aden Airways.
1553	**—**	Srs.2: D. Brazilian Air Force.
1554	**—**	Srs.2: D. Brazilian Air Force.
1555	**—**	Srs.2: D. Brazilian Air Force.
1556	**—**	Srs.2: D. Brazilian Air Force.

BOEING 707

17586	**N708PA**	707-121: F.F. 20.12.57. D. PAA 1.12.58 Constitution.
17587	**N707PA**	707-121: D. PAA 19.12.58. Maria.
17588	**N709PA**	707-121: D. PAA 15.8.58. Tradewind.
17589	**N710PA**	707-121: D. PAA 26.9.58. America.
17590	**N711PA**	707-121: D. PAA 16.10.58. Mayflower. L. Avianca 1961.
17591	**N712PA**	707-121: D. PAA 31.10.58. Washington.
17592	**N714PA**	707-321: D. PAA 28.8.59. Golden Eagle.
17593	**N715PA**	707-321: D. PAA 19.7.59. Liberty Bell.
17594	**N716PA**	707-321: D. PAA 22.8.59. Flying Eagle.
17595	**N717PA**	707-321: D. PAA 1.9.59. Fleetwing.
17596	**N718PA**	707-321: D. PAA 22.9.59. Invincible.
17597	**N719PA**	707-321: D. PAA 2.10.59. Windward.
17598	**N720PA**	707-321: D. PAA 6.10.59. Fairwind.
17599	**N721PA**	707-321: D. PAA 19.10.59. Splendid.
17600	**N722PA**	707-321: D. PAA 29.10.59. Lark.
17601	**N723PA**	707-321: D. PAA 27.10.59. Viking. L. PIA 1960–61.
17602	**N724PA**	707-321: D. PAA 9.12.59. Mercury.
17603	**N725PA**	707-321: D. PAA 13.12.59. Aurora.
17604	**N726PA**	707-321: D. PAA 13.1.60. Westward Ho.
17605	**N727PA**	707-321: D. PAA 28.1.60. Mohawk.
17606	**N728PA**	707-321: D. PAA 5.3.60. Peerless.
17607	**N729PA**	707-321: D. PAA 26.4.60. Isabella.
17608	**N730PA**	707-321: D. PAA 28.4.60. Bald Eagle.
17609	**N70773**	707-124: F.F. 25.3.59. D. Continental 19.4.59.
17610	**N70774**	707-124: D. Continental 28.5.59.
17611	**N70775**	707-124: D. Continental 16.7.59.
17612	**N70785**	707-124: D. Continental 10.8.59.
17613	**F-BHSA**	707-328: F.F. 11.9.59. D. Air France 6.11.59. Versailles. W.O. 27.6.61.
17614	**F-BHSB**	707-328: D. Air France 11.12.59. Chambord.
17615	**F-BHSC**	707-328: F.F. as N74615. D. Air France 22.12.59. Fontainebleau.
17616	**F-BHSD**	707-328: D. Air France 29.1.60. Chenonceaux.
17617	**F-BHSE**	707-328: D. Air France 18.3.60. Rambouillet.
17618	**F-BHSF**	707-328: D. Air France 24.3.60. Blois.
17619	**F-BHSG**	707-328: F.F. as N5093K. D. Air France 12.5.60. Pau.
17620	**F-BHSH**	707-328: D. Air France 23.6.60. Amboise.
17621	**F-BHSI**	707-328: D. Air France 11.7.60. Josselin.
17622	**F-BHSJ**	707-328: D. Air France 29.7.60. Chaumont.
17623	**OO-SJA**	707-329: F.F. 3.11.59. D. Sabena 4.12.59.
17624	**OO-SJB**	707-329: D. Sabena 14.1.60. W.O. 15.2.61.
17625	**OO-SJC**	707-329: D. Sabena 12.2.60.
17626	**OO-SJD**	707-329: D. Sabena 9.4.60.
17627	**OO-SJE**	707-329: D. Sabena 7.6.60.
17628	**N7501A**	707-123B: F.F. 5.10.58. D. American Airlines 23.10.58. Michigan.
17629	**N7502A**	707-123: D. American Airlines 4.5.59. Oklahoma. W.O. 28.1.61.
17630	**N7503A**	707-123B: D. American Airlines 31.12.58. California.
17631	**N7504A**	707-123B: D. American Airlines 28.1.59. New York.
17632	**N7505A**	707-123B: D. American Airlines 31.1.59. Illinois.
17633	**N7506A**	707-123B: D. American Airlines 13.2.59. District of Columbia.
17634	**N7507A**	707-123B: D. American Airlines 28.2.59. Maryland.
17635	**N7508A**	707-123B: D. American Airlines 27.3.59. Virginia. W.O. 1.3.62.
17636	**N7509A**	707-123B: D. American Airlines 9.4.59. Texas.
17637	**N7510A**	707-123B: D. American Airlines 23.4.59. Massachusetts.
17638	**N7511A**	707-123B: D. American Airlines 12.5.59. New Jersey.
17639	**N7512A**	707-123B: D. American Airlines 21.5.59. Pennsylvania.
17640	**N7513A**	707-123B: D. American Airlines 28.5.59. Delaware.
17641	**N7514A**	707-123: D. American Airlines 5.6.59. Connecticut. W.O. 15.8.59.

17642	N7515A	707-123B: *D.* American Airlines 24.6.59. *Arizona.*
17643	N7516A	707-123B: *D.* American Airlines 29.6.59. *Ohio.*
17644	N7517A	707-123B: *D.* American Airlines 27.7.59. *Missouri.*
17645	N7518A	707-123B: *D.* American Airlines 31.7.59. *Tennessee.*
17646	N7519A	707-123B: *D.* American Airlines 22.8.59. *Kentucky.*
17647	N7520A	707-123B: *D.* American Airlines 12.8.59. *Arkansas.*
17648	N7521A	707-123B: *D.* American Airlines 15.9.59. *West Virginia.*
17649	N7522A	707-123B: *D.* American Airlines 5.10.59. *Indiana.*
17650	N7523A	707-123B: *D.* American Airlines 14.10.59. *Rhode Island.*
17651	N7524A	707-123B: *D.* American Airlines 28.10.59. *Canada.*
17652	N7525A	707-123B: *D.* American Airlines 20.11.59. *Mexico.*
17658	N731TW	707-131: *F.F.* 3.12.58. *D.* TWA 29.1.59.
17659	N732TW	707-131: *D.* TWA 17.3.59.
17660	N733TW	707-131: *D.* TWA 30.3.59.
17661	N734TW	707-131: *D.* TWA 3.4.59.
17662	N735TW	707-131: *D.* TWA 18.4.59.
17663	N736TW	707-131: *D.* TWA 29.4.59.
17664	N737TW	707-131: *D.* TWA 10.5.59.
17665	N738TW	707-131: *D.* TWA 13.5.59.
17666	N739TW	707-131: *D.* TWA 24.5.59.
17667	N740TW	707-131: *D.* TWA 28.5.59.
17668	N741TW	707-131: *D.* TWA 13.6.59.
17669	N742TW	707-131: *D.* TWA 1.7.59.
17670	N743TW	707-131: *D.* TWA 10.7.59.
17671	N744TW	707-131: *D.* TWA 14.7.59.
17672	N745TW	707-131: *D.* TWA 1.8.59.
17673	N761TW	707-331: *D.* TWA 10.11.59.
17674	N701PA	707-331: *D.* PAA 6.11.59. *Donald McKay.*
17675	N762TW	707-331: *D.* TWA 10.11.59.
17676	N763TW	707-331: *D.* TWA 25.11.59.
17677	N702PA	707-331: *D.* PAA 15.12.59. *Hotspur.*
17678	N764TW	707-331: *D.* TWA 23.12.59.
17679	N765TW	707-331: *D.* TWA 18.1.60.
17680	N703PA	707-331: *D.* PAA 30.12.59. *Dashaway.*
17681	N766TW	707-331: *D.* TWA 1.4.60.
17682	N767TW	707-331: *D.* TWA 5.4.60.
17683	N704PA	707-331: *D.* PAA 23.3.60. *Defiance.*
17684	N768TW	707-331: *D.* TWA 14.4.60.
17685	N769TW	707-331: *D.* TWA 9.5.60.
17686	N705PA	707-331: *D.* PAA 29.4.60. *Wings of the Morning.*
17687	N770TW	707-331: *D.* TWA 25.5.60.
17688	N771TW	707-331: *D.* TWA 1.7.60.
17689	N706PA	707-331: *D.* PAA 8.6.60. *Courier.*
17690	N772TW	707-331: *D.* TWA 1.7.60.
17691	N7071	707-227: *F.F.* 11.6.59 (for Braniff). *W.O.* 19.10.59.
17692	N7072	707-227: *D.* Braniff 3.12.59.
17693	N7073	707-227: *D.* Braniff 15.1.60.
17694	N7074	707-227: *D.* Braniff 21.1.60.
17695	N7075	707-227: *D.* Braniff 10.2.60.
17696	VH-EBA	707-138B: *F.F.* as N31239. 20.3.59. *D.* Qantas 16.7.59. *City of Melbourne.*
17697	VH-EBB	707-138B: *D.* Qantas 26.6.59. *City of Sydney.*
17698	VH-EBC	707-138B: *D.* Qantas 6.7.59. *City of Canberra.*
17699	VH-EBD	707-138B: *D.* Qantas 1.8.59. *City of Brisbane.*
17700	VH-EBE	707-138B: *D.* Qantas 24.8.59. *City of Perth.*
17701	VH-EBF	707-138B: *D.* Qantas 4.9.59. *City of Adelaide.*
17702	VH-EBG	707-138B: *D.* Qantas 18.9.59. *City of Hobart.*
17703	G-APFB	707-436: *F.F.* as N31241, 19.5.59. *D.* BOAC 9.5.60.
17704	G-APFC	707-436: *F.F.* as N5088K. *D.* BOAC 16.5.60.
17705	G-APFD	707-436: *F.F.* as N5091K. *D.* BOAC 27.4.60.
17706	G-APFE	707-436: *F.F.* as N5092K. *D.* BOAC 29.4.60.
17707	G-APFF	707-436: *D.* BOAC 13.5.60.
17708	G-APFG	707-436: *F.F.* as N5094K. *D.* BOAC 22.6.60.
17709	G-APFH	707-436: *D.* BOAC 13.7.60.
17710	G-APFI	707-436: *D.* BOAC 23.7.60.
17711	G-APFJ	707-436: *D.* BOAC 22.9.60.
17712	G-APFK	707-436: *D.* BOAC 28.9.60.
17713	G-APFL	707-436: *D.* BOAC 21.10.60.
17714	G-APFM	707-436: *D.* BOAC 4.11.60.
17715	G-APFN	707-436: *D.* BOAC 16.11.60.
17716	G-APFO	707-436: *D.* BOAC 9.12.60.
17717	G-APFP	707-436: *D.* BOAC 22.12.60.
17718	D-ABOB	707-430: *F.F.* as N31240. 18.12.59. *D.* Lufthansa 25.2.60. *Hamburg.*
17719	D-ABOC	707-430: *D.* Lufthansa 10.3.60. *Berlin.*
17720	D-ABOD	707-430: *D.* Lufthansa 23.4.60. *Frankfurt.*
17721	D-ABOF	707-430: *D.* Lufthansa 16.3.61. *München.*
17722	VT-DJI	707-437: *F.F.* 14.1.60. *D.* Air India 18.2.60. *Nanda Devi.*
17723	VT-DJJ	707-437: *D.* Air India 18.2.60. *Annapurna.*
17724	VT-DJK	707-437: *D.* Air India 7.3.60. *Everest.*
17903	N74613	707-139: *F.F.* 20.3.60. *Ordered* Cubana. *D.* Western (L.) 4.5.60.
17904	N74614	707-139: *Ordered* Cubana. *D.* Western (L.) 13.5.60.
17905	PP-VJA	707-441: *F.F.* as N5090K. 4.2.60. *D.* Varig 7.6.60.
17906	PP-VJB	707-441: *D.* Varig 15.6.60.
17907	N7201U	720-022: *F.F.* 23.11.59. *D.* United 1.10.60.
17908	N7202U	720-022: *D.* United 29.7.60.
17909	N7203U	720-022: *D.* United 30.4.60.
17910	N7204U	720-022: *D.* United 21.5.60. *Philip G. Johnson.*
17911	N7205U	720-022: *D* United 25.5.60.
17912	N7206U	720-022: *D.* United 9.6.60.
17913	N7207U	720-022: *D.* United 29.6.60.
17914	N7208U	720-022: *D.* United 25.6.60.
17915	N7209U	720-022: *D.* United 27.7.60.
17916	N7210U	720-022: *D.* United 5.8.60.
17917	N7211U	720-022: *D.* United 13.8.60.
17918	F-BHSK	707-328: *F.F.* as N5095K. *D.* Air France 19.8.60. *Vizille.*
17919	F-BHSL	707-328: *D.* Air France 19.8.60. *Maintenon.*
17920	F-BHSM	707-328: *D.* Air France 20.9.60. *Sully.*
17921	F-BHSN	707-328: *D.* Air France 18.9.60. *Valencay.*
17922	F-BHSO	707-328: *D.* Air France 16.9.60. *Anet.*
17923	F-BHSP	707-328: *D.* Air France 7.11.60. *Villandry.*
17924	F-BHSQ	707-328: *D.* Air France 22.10.60. *Compeigne.*
17925	58-6970	707-153: *F.F.* 4.4.59. *D.* USAF 4.5.59 as VC-137A.
17926	58-6971	707-153: *D.* USAF 31.5.59 as VC-137A.
17927	58-6972	707-153: *D.* USAF 30.6.59 as VC-137A.
17928	ZS-CKC	707-344: *F.F.* 25.5.60. *D.* SAA 1.7.60. *Johannesburg.*
17929	ZS-CKD	707-344: *D.* SAA 22.8.60. *Capetown.*
17930	ZS-CKE	707-344: *D.* SAA 22.8.60. *Durban.*
18012	N74612	707-124: *D.* Continental 1962.
18013	N7527A	720-023B: *D.* American 30.7.60. *Mississippi.*
18014	N7528A	720-023B: *D.* American 24.7.60. *Connecticut.*
18015	N7529A	720-023B: *D.* American 13.8.60. *Wisconsin.*
18016	N7530A	720-023B: *D.* American 22.9.60. *Alabama.*
18017	N7531A	720-023B: *D.* American 1.9.60. *North Dakota.*
18018	N7532A	720-023B: *D.* American 8.9.60. *South Carolina.*
18019	N7533A	720-023B: *D.* American Airlines 19.9.60. *South Dakota.*
18020	N7534A	720-023B: *D.* American Airlines 10.10.60. *Colorado.*
18021	N7535A	720-023B: *D.* American Airlines 23.11.60. *Florida.*
18022	N7536A	720-023B: *D.* American Airlines 3.12.60. *Georgia.*
18023	N7537A	720-023B: *D.* American Airlines 27.4.61.
18024	N7538A	720-023B: *D.* American Airlines 3.2.61.
18025	N7539A	720-023B: *D.* American Airlines 17.3.61.
18026	N7540A	720-023B: *D.* American Airlines 17.2.61.

BOEING 707—*continued*

18027	N7541A	720-023B: *D.* American Airlines 27.2.61.
18028	N7542A	720-023B: *D.* American Airlines 29.3.61. *Maine.*
18029	N7543A	720-023B: *D.* American Airlines 28.3.61. *Minnesota.*
18030	N7544A	720-023B: *D.* American Airlines 10.4.61.
18031	N7545A	720-023B: *D.* American Airlines 17.4.61.
18032	N7546A	720-023B: *D.* American Airlines 20.4.61.
18033	N7547A	720-023B: *D.* American Airlines 23.5.61.
18034	N7548A	720-023B: *D.* American Airlines 22.6.61.
18035	N7549A	720-023B: *D.* American Airlines 9.6.61.
18036	N7550A	720-023B: *D.* American Airlines 1.7.61.
18037	N7551A	720-023B: *D.* American Airlines 20.7.61.
18041	EI-ALA	720-048: *D.* Irish Airlines 24.10.60. *St. Patrick.*
18042	EI-ALB	720-048: *D.* Irish Airlines 24.1.61. *St. Briget.*
18043	EI-ALC	720-048: *D.* Irish Airlines 6.4.61. *St. Brendan.*
18044	N7212U	720-022: *D.* United 22.12.60.
18045	N7213U	720-022: *D.* United 19.12.60.
18046	N7214U	720-022: *D.* United 21.1.61.
18047	N7215U	720-022: *D.* United 27.1.61.
18048	N7216U	720-022: *D.* United 2.2.61.
18049	N7217U	720-022: *D.* United 13.2.61.
18050	N7218U	720-022: *D.* United 5.3.61.
18054	N7526A	707-123B: *D.* American 25.5.61.
18055	VT-DMN	707-437: *D.* Air India 17.4.61. *Kanchenjunga.*
18056	D-ABOG	707-430: *D.* Lufthansa 16.3.61. *Bonn.*
18057	D-ABOH	720-030B: *D.* Lufthansa 8.3.61. *Köln.*
18058	D-ABOK	720-030B: *D.* Lufthansa 28.4.61. *W.O.* 4.12.61.
18059	D-ABOL	720-030B: *D.* Lufthansa 3.5.61. *Stuttgart.*
18060	D-ABOM	720-030B: *D.* Lufthansa 3.6.61. *Nürnberg.*
18061	N93141	720-047B: *D.* Western 7.4.61.
18062	N93142	720-047B: *D.* Western 9.5.61.
18063	N93143	720-047B: *D.* Western 7.6.61.
18064	N7076	720-027: *D.* Braniff 11.2.61.
18065	N7077	720-027: *D.* Braniff 22.3.61.
18066	N113	720-061: *D.* FAA 12.5.61.
18067	VH-EBH	707-138B: *F.F.* as N93134. *D.* Qantas 29.7.61. *City of Darwin.*
18068	VH-EBI	707-138B: *D.* Qantas 16.8.61. *Winton.*
18069	VH-EBJ	707-138B: *D.* Qantas 24.8.61. *Longreach.*
18070	4X-ATA	707-458: *D.* El Al. 24.4.61. *Shehecheyanu.*
18071	4X-ATB	707-458: *D.* El Al. 8.6.61.
18072	N7219U	720-022: *D.* United 1.12.61.
18073	N7220U	720-022: *D.* United 13.12.61.
18074	N7221U	720-022: *D.* United 21.12.61.
18075	N7222U	720-022: *D.* United 10.1.62.
18076	N7223U	720-022: *D.* United 17.1.62.
18077	N7224U	720-022: *D.* United.
18078	N7225U	720-022: *D.* United.
18079	N7226U	720-022: *D.* United.
18080	N7227U	720-022: *D.* United.
18081	N7228U	720-022: *D.* United.
18082	N7229U	720-022: *D.* United.
18083	N757PA	707-321: *D.* PAA 16.5.61. *Pathfinder.*
18084	N758PA	707-321: *D.* PAA 23.5.61. *Resolute.*
18085	N759PA	707-321: *D.* PAA 13.6.61. *Freedom.*
18086	HK724	720-059B: *D.* Avianca 8.11.61.
18087	HK725	720-059B: *D.* Avianca 15.11.61.
18154	N7078	720-027: *D.* Braniff 9.8.61.
18155	N8701E	720-025: *D.* Eastern 11.8.61.
18156	N8702E	720-025: *D.* Eastern 25.8.61.
18157	N8703E	720-025: *D.* Eastern 2.2.62.
18158	N8704E	720-025: *D.* Eastern 19.9.61.
18159	N8705E	720-025: *D.* Eastern 27.9.61.
18160	N8706E	720-025: *D.* Eastern 6.10.61.
18161	N8707E	720-025: *D.* Eastern 17.10.61.
18162	N8708E	720-025: *D.* Eastern 8.11.61.
18163	N8709E	720-025: *D.* Eastern 23.10.61.
18164	N8710E	720-025: *D.* Eastern 23.10.61.
18165	HZ-ACA	720-068B: *D.* Saudi Arabia 20.12.61.
18166	HZ-ACB	720-068B: *D.* Saudi Arabia 20.12.61.
18167	N93144	720-047B: *D.* Western 11.7.61.
18240	N8711E	720-025: *D.* Eastern 8.1.62.
18241	N8712E	720-025: *D.* Eastern 13.11.61.
18242	N8713E	720-025: *D.* Eastern 22.11.61.
18243	N8714E	720-025: *D.* Eastern 8.12.61.
18244	N8715E	720-025: *D.* Eastern 16.12.61.
18245	F-BHSR	707-328: *F.F.* as N93138. *D.* Air France 31.1.62. *Chantilly.*
18246	F-BHSS	707-328: *D.* Air France 16.2.62. *d'Uzes.*
18247	F-BHST	707-328: *D.* Air France 9.3.62. *Cheverny.*
18248	D-ABON	720-030B: *D.* Lufthansa 5.1.62. *Hanover.*
18249	D-ABOP	720-030B: *D.* Lufthansa 12.1.62.
18250	D-ABOQ	720-030B: *F.F.* as N93137. *D.* Lufthansa.
18251	D-ABOR	720-030B: *D.* Lufthansa 27.2.62. *Dortmund.*
18334	VH-EBK	707-138B: *D.* Qantas 29.8.61. *City of Newcastle.*
18335	N760PA	707-321B: *F.F.* 31.1.62. *D.* PAA 1962. *Evening Star.*
18336	N761PA	707-321B: *F.F.* 26.2.62. *D.* PAA 1962. *Friendship.*
18337	N762PA	707-321B: *D.* PAA 16.4.62. *Endeavour.*
18338	N763PA	707-321B: *D.* PAA 1962. *Yankee.*
18339	N764PA	707-321B: *D.* PAA 1962. *Nautilus.*
18351	N721US	720-051B: *D.* Northwest 26.5.61.
18352	N722US	720-051B: *D.* Northwest 22.6.61.
18353	N723US	720-051B: *D.* Northwest 12.7.61.
18354	N724US	720-051B: *D.* Northwest 26.7.61.
18355	N725US	720-051B: *D.* Northwest 1.9.61.
18356	N726US	720-051B: *D.* Northwest 5.10.61.
18357	4X-ATC	707-458: *D.* El Al 15.2.62.
18372	VR-BBW	707-465: *D.* Cunard Eagle 27.2.62 as G-ARWD.
18373	—	707-465: *D.* Cunard Eagle.
18374	OO-SJF	707-329: *D.* Sabena 17.4.62.
18375	F-BHSU	707-328: *D.* Air France.
18376	N720V	720-062: *D.* Pacific Northern.
18377	N720W	720-062: *D.* Pacific Northern.
18378	AP-AMG	720-040B: *D.* PIA 26.12.61.
18379	AP-AMH	720-040B: *D.* PIA.
18380	AP-AMJ	720-040B: *D.* PIA.
18381	N791TW	720-051B: *D.* TWA 23.7.61.
18382	N792TW	720-051B: *D.* TWA 2.8.61.
18383	N793TW	720-051B: *D.* TWA 27.8.61.
18384	N795TW	720-051B: *D.* TWA 30.9.61.
18385	N746TW	707-131B: *D.* TWA 3.62.
18386	N747TW	707-131B: *D.* TWA 4.62.
18387	N748TW	707-131B: *D.* TWA.
18388	N749TW	707-131B: *D.* TWA.
18389	N750TW	707-131B: *D.* TWA.
18390	N751TW	707-131B: *D.* TWA.
18391	N752TW	707-131B: *D.* TWA.
18392	N754TW	707-131B: *D.* TWA.
18393	N755TW	707-131B: *D.* TWA.
18394	N756TW	707-131B: *D.* TWA.
18395	N757TW	707-131B: *D.* TWA.
18396	N758TW	707-131B: *D.* TWA.
18397	N759TW	707-131B: *D.* TWA.
18398	N760TW	707-131B: *D.* TWA.

18399 N761TW 707-131B: *D.* TWA.
18400 N781TW 707-131B: *D.* TWA.
18401 N782TW 707-131B: *D.* TWA.
18402 N783TW 707-131B: *D.* TWA.
18405 N773TW 707-331B: *D.* TWA (L.).
18406 N774TW 707-331B: *D.* TWA (L.).
18407 N775TW 707-331B: *D.* TWA (L.).
18408 N776TW 707-331B: *D.* TWA (L.).
18409 N778TW 707-331B: *D.* TWA (L.).
18411 G-ARRA 707-436: *D.* BOAC 16.2.62.
18412 G-ARRB 707-436: *D.* BOAC 1963.
18413 G-ARRC 707-436: *D.* BOAC 1963.
18414 VT-DNY 707-437: *D.* Air India 7.3.62. *Dhaulagiri.*
18415 VT-DNZ 707-437: *D.* Air India 4.62. *Nangaparbat.*
18416 N57201 720-024B: *D.* Continental 30.4.62.
18417 N57202 720-024B: *D.* Continental.
18418 N57203 720-024B: *D.* Continental.
18419 N57204 720-024B: *D.* Continental.
18420 N727US 720-051B: *D.* Northwest 25.10.61.
18421 N728US 720-051B: *D.* Northwest (L.) 15.11.61.
18422 N729US 720-051B: *D.* Northwest (L.) 13.12.61.
18423 N7079 720-027: *D.* Braniff.
18424 4X-ABA 720-058B: *D.* El Al 3.62.
18425 4X-ABB 720-058B: *D.* El Al 26.4.62.
18451 N93145 720-047B: *D.* Western.
18452 N93146 720-047B: *D.* Western.
18453 N93147 720-047B: *D.* Western.
18454 ET-AAG 720-060B: *D.* Ethiopian Airlines.
18455 ET-AAH 720-060B: *D.* Ethiopian Airlines.
18456 F- 707-328B: *D.* Air France 1963. *Vincennes.*
18457 F- 707-328B: *D.* Air France 1963. *Luneville.*
18458 F- 707-328B: *D.* Air France 1963. *Kerjean.*
18459 F- 707-328B: *D.* Air France 1963. *Grignan.*
18460 OO-SJG 707-329: *D.* Sabena 1963.
18461 62-6000 707-353B: *D.* USAF as VC-137C 1963.
18462 D-ABOS 707-330B: *D.* Lufthansa 1963.
18463 D-ABOT 707-330B: *D.* Lufthansa 1963.

BRISTOL BRITANNIA

12873 G-ALBO Proto.101: *F.F.* 16.8.52. *Last F.* 30.11.60, to RAF St. Athan as 7708M.
12874 G-ALRX 2nd Proto.101: *F.F.* 23.12.53. *W.O.* 4.2.54.
12902 G-ANBA Srs.102: *F.F.* 5.9.54. *D.* BOAC 12.8.57.
12903 G-ANBB Srs.102: *F.F.* 18.1.55. *D.* BOAC 10.6.57.
12904 G-ANBC Srs.102: *F.F.* 29.6.55. *D.* BOAC 30.12.55. *W.O.* 11.11.60.
12905 G-ANBD Srs.102: *F.F.* 14.11.55. *D.* BOAC 30.12.55.
12906 G-ANBE Srs.102: *F.F.* 17.1.56. *D.* BOAC 2.3.56.
12907 G-ANBF Srs.102: *F.F.* 23.2.56. *D.* BOAC 14.3.56.
12908 G-APLL Srs.102: *F.F.* as G-ANBG 29.3.56. *D.* BOAC 8.8.56.
12909 G-ANBH Srs.102: *F.F.* 9.5.56. *D.* BOAC 24.7.57.
12910 G-ANBI Srs.102: *F.F.* 24. 5.56. *D.* BOAC 29.6.56.
12911 G-ANBJ Srs.102: *F.F.* 5.8.56. *D.* BOAC 22.11.56.
12912 G-ANBK Srs.102: *F.F.* 14.9.56. *D.* BOAC 28.11.56.
12913 G-ANBL Srs.102: *F.F.* 24.2.57. *D.* BOAC 2.3.57.
12914 G-ANBM Srs.102: *F.F.* 6.3.57 *D.* BOAC 11.3.57.
12915 G-ANBN Srs.102: *F.F.* 11.4.57. *D.* BOAC 4.5.57.
12916 G-ANBO Srs.102: *F.F.* 17.5.57. *D.* BOAC 31.5.57. *L.* Malayan Airways 1962.
12917 G-ANCA Proto. 301: *F.F.* 31.7.56. *W.O.* 6.11.57.
12918 XA-MEC Srs.302: ordered BOAC as G-ANCB. *F.F.* as G-18-1, 21.6.57. *D.* Aeronaves 5.11.57. *Tenochtitlan.*

12919 XA-MED Srs.302: ordered BOAC as G-ANCC. *F.F.* as G-18-2, 24.7.57. *D.* Aeronaves 15.12.57. *Tzintzuntzan.*
12920 G-ANCD Srs.305: *F.F.* as G-18-3, 1.6.57. Assigned Northeast Airlines as N6595C, not delivered. Leased El Al as 4X-AGE, 17.7.58 to 6.3.59 as Srs.306. To Air Charter (now BUA) 24.3.59 as Srs.307.
12921 G-ANCE Srs.305: Assigned Northeast Airlines as N6596C, not delivered. *F.F.* 3.9.58 as Srs.307. *D.* Air Charter (now BUA) 12.9.58.
12922 LV-GJB Srs.305: Assigned Northeast Airlines as N6597C, not delivered. *F.F.* 19.11.58 as G-18-4. *Cvtd.* Type 308 G-14-1. *D.* Transcontinental S.A. as LV-PPJ, 16.12.59.
12923 LV-GJC Srs.305: Assigned Northeast Airlines as N6598C, not delivered. *Cvtd.* to Srs.308 and *D.* Transcontinental S.A. as LV-PPL, 17.12.59.
12924 9G-AAG Srs.305: ordered BOAC allocated Northeast as N6599C. *F.F.* as G-ANCH, 19.2.60. *Cvtd.* Srs.309 and *D.* Ghana Airways 27.7.60.
12925 G-AOVH Originally Srs.305 G-ANCI. Built as Srs.312. *F.F.* 29.1.58. *D.* BOAC 11.2.58.
12926 G-AOVI Originally Srs.305 G-ANCJ. Built as Srs.312. *F.F.* 15.2.58. *D.* BOAC 10.4.58.
13207 9G-AAH Srs.311: *F.F.* 31.12.56 as G-AOVA. *D.* Ghana Airways 8.11.60 as Srs.319 *Osagyefo.*
13230 G-AOVB Srs.312: *F.F.* 5.7.57. *D.* BOAC 10.9.57.
13231 G-AOVC Srs.312: *F.F.* 22.10.57. *D.* BOAC 15.11.57.
13232 4X-AGA Srs.313: *F.F.* 28.7.57. *D.* El Al 12.9.57.
13233 G-ARWX Srs.313: *F.F.* 2.9.57. *D.* El Al 19.10.57 as 4X-AGB. *S.* BUA, 2.62.
13234 G-ARXA Srs.313: *F.F.* 4.10.57. *D.* El Al 28.11.57 as 4X-AGC. *S.* BUA, 2.62.
13235 G-AOVD Srs.312: *F.F.* 13.11.57. *D.* BOAC 6.12.57. *W.O.* 24.12.58.
13236 G-AOVE Srs.312: *F.F.* 8.12.57. *D.* BOAC 21.12.57. *L.* BUA 5.10.61.
13237 G-AOVF Srs.312: *F.F.* 18.12.57. *D.* BOAC 2.1.58.
13238 G-AOVG Srs.312: *F.F.* 10.1.58. *D.* BOAC 19.3.58.
13393 CF-CZA Srs.314: *F.F.* 11.1.58. *D.* CPAL 9.4.58. *Buenos Aires* later *Hong Kong.*
13394 CF-CZB Srs.314: *F.F.* 14.4.58. *D.* CPAL 29.4.58. *Vancouver.*
13395 CF-CZC Srs.314: *F.F.* 13.5.58. *D.* CPAL 20.5.58 *Madrid.*
13396 CF-CZD Srs.314: *F.F.* 13.6.58. *D.* CPAL 27.6.58 *Santiago,* later *Winnipeg.*
13397 XL635 Srs.253: C.Mk.1. *F.F.* 29.12.59. *D.* 29.1.60.
13398 XL636 Srs.253: C.Mk.1. *F.F.* 23.4.59. *D.* No. 99 Sqdn. 4.6.59 *Argo.*
13399 XL637 Srs.253: C.Mk.1. *F.F.* 12.5.59. *D.* No. 99 Sqdn. 26.6.59 *Vecia.*
13400 XL638 Srs.253: C.Mk.1. *F.F.* 22.6.59. *D.* No. 99 Sqdn. 5.8.59 *Sirius.*
13418 G-AOVJ Srs.312: *F.F.* 27.2.58. *D.* BOAC 13.3.58.
13419 G-AOVK Srs.312: *F.F.* 18.3.58. *D.* BOAC 5.58.
13420 G-AOVL Srs.312: *F.F.* 9.4.58. *D.* BOAC 5.58.
13421 G-AOVM Srs.312: *F.F.* 29.4.58. *D.* BOAC 10.6.58.
13422 G-AOVN Srs.312: *F.F.* 16.5.58. *D.* BOAC 4.7.58.
13423 G-AOVO Srs.312: *F.F.* 3.7.58. *D.* BOAC 4.9.58.
13424 G-AOVP Srs.312: *F.F.* 22.7.58. *D.* BOAC 31.7.58.
13425 G-APNA Srs.317: *F.F.* 10.10.58. *D.* Hunting Clan (BUA) 11.12.58.
13426 G-APNB Srs.317: *F.F.* 10.11.58. *D.* Hunting Clan (BUA) 2.12.58.
13427 G-AOVT Srs.312: *F.F.* 17.12.58. *D.* BOAC 1.1.59.
13428 CF-CZX Srs.314: *F.F.* 19.6.58. *D.* CPAL 3.7.58 *Santa Maria,* later *Canada.*
13429 G-AOVR Srs.312: *F.F.* 4.8.58. *D.* BOAC 3.10.58.
13430 G-AOVS Srs.312: *F.F.* 5.9.58. *D.* BOAC 39.10.58.
13431 4X-AGD Srs.313: *F.F.* 21.2.59. *D.* El Al 7.3.59.
13432 OK-MBA Srs.318: *F.F.* 24.11.58 as CU-P668. *D.* Cubana 15.12.58, re-reg. CU-T668 *Liberated.* *L.* Eagle (Cunard Eagle) 24.3.60 to 19.9.61 as G-APYY. *L.* CSA 10.61.
13433 CU-T669 Srs.318: *F.F.* 19.1.59 as CU-P669. *D.* Cubana 6.2.59.
13434 XM489 Srs.253: C.Mk.1. *F.F.* 7.4.60. *D.* No. 511 Sqdn. 4.5.60. *Denebola.*
13435 XM490 Srs.253: C.Mk.1. *F.F.* 16.5.60. *D.* No. 511 Sqdn. 8.6.60 *Aldebaran.*
13436 XM491 Srs.253: C.Mk.1. *F.F.* 16.6.60. *D.* No. 511 Sqdn. 6.7.60 *Procyon.*
13437 CU-T670 Srs.318/2: *F.F.* 2.4.59 as CU-P670. *D.* Cubana 15.5.59.
13448 XL639 Srs.253: C.Mk.1. *F.F.* 28.8.59. *D.* No. 99 Sqdn. 7.10. 59 *Atria.*
13449 XL640 Srs.253: C.Mk.1. *F.F.* 8.10.59. *D.* No. 511 Sqdn. 30.10.59 *Antares.*

BRISTOL BRITANNIA—*continued*

13450 **XN392** Srs.252: *F.F.* 13.10.58 as G-APPE. *D.* No. 99 Sqdn. as C.Mk.2, 27.10.59 *Accrux.*
13451 **XN398** Srs.252: *F.F.* 7.12.58 as G-APPF. *D.* No. 99 Sqdn. as C.Mk.2, 19.3.59 *Altair.*
13452 **XN404** Srs.252: *F.F.* 3.3.59 as G-APPG. *D.* No. 99 Sqdn. as C.Mk.2, 8.4.59 *Canopus.*
13453 **CF-CZW** Srs.314: *F.F.* 22.7.58. *D.* CPA 7.8.58 *Edmonton,* later *Toronto.*
13454 **XL657** Srs.253: C.Mk.1. *F.F.* 23.11.59. *D.* No. 511 Sqdn. 23.12.59 *Rigel.*
13455 **XL658** Srs.253: C.Mk.1. *F.F.* 3.12.59. *D.* No. 511 Sqdn. 4.2.60 *Adhara.*
13456 **XL659** Srs.253: C.Mk.1. *F.F.* 1.2.60. *D.* No. 511 Sqdn. 4.3.60 *Polaris.*
13457 **XL660** Srs.253: C.Mk.1. *F.F.* 4.3.60. *D.* No. 99 Sqdn. 23.4.60 *Alphard.*
13508 **XM496** Srs.253: C.Mk.1. *F.F.* 24.8.60. *D.* No. 99 Sqdn. 17.9.60 *Regulus.*
13509 **XM497** Srs.253: C.Mk.1. *F.F.* 17.11.60. *D.* Lyneham 2.12.60 *Schedar.*
13510 **XM498** Srs.253: C.Mk.1. *F.F.* 30.9.59. *D.* No. 99 Sqdn. 17.10.59 *Hadar.*
13511 **XM517** Srs.253: C.Mk.1. *F.F.* 24.11.59. *D.* Lyneham 12.59.
13512 **XM518** Srs.253: C.Mk.1. *F.F.* 18.12.59. *D.* No. 511 Sqdn. 19.12.59 *Spica.*
13513 **XM519** Srs.253: C.Mk.1. *F.F.* 28.1.60. *D.* No. 511 Sqdn. 5.2.56 *Capella.*
13514 **XM520** Srs.253: C.Mk.1. *F.F.* 9.3.60. *D.* No. 99 Sqdn. 19.3.60 *Arcturus.*
13515 **CU-T671** Srs.318/2: *F.F.* 29.4.59 as CU-P671. *D.* Cubana 22.8.59.
13516 **G-ARKA** Srs.324: *F.F.* 9.10.59. *D.* CPAL as CF-CPO, 16.10.59. *Amsterdam. S.* Cunard Eagle, 8.3.61.
13517 **G-ARKB** Srs.324: *F.F.* 4.11.59. *D.* CPAL as CF-CPE, 13.11.59. *Mexico City. S.* Cunard Eagle, 2.5.61.

CANADAIR FORTY FOUR

1 **15555** CL-44-6 (CC-108) *F.F.* 15.11.59 as 15501 *Cvtd.* to VIP.
2 **16666** CL-44-6 (CC-108) *F.F.* as 15922. *Cvtd.* to VIP.
3 **15923** CL-44-6 (CC-108) RCAF.
4 **15924** CL-44-6 (CC-108) RCAF.
5 **15925** CL-44-6 (CC-108) RCAF.
6 **15926** CL-44-6 (CC-108) RCAF.
7 **15927** CL-44-6 (CC-108) RCAF.
8 **15928** CL-44-6 (CC-108) RCAF.
9 **CF-MKP-X** CL-44D-4 Demonstrator *F.F.* 16.11.60.
10 **15929** CL-44-6 (CC-108) RCAF.
11 **15930** CL-44-6 (CC-108) RCAF.
12 **15931** CL-44-6 (CC-108) RCAF.
13 **15932** CL-44-6 (CC-108) RCAF.
14 **N124SW** CL-44D-4 *D.* Seaboard 20.6.61.
15 **N446T** CL-44D-4 *D.* FTL.
16 **N447T** CL-44D-4 *D.* FTL. 31.5.61.
17 **N448T** CL-44D-4 *D.* FTL.
18 **N449T** CL-44D-4 *D.* FTL.
19 **N450T** CL-44D-4 *D.* FTL. 10.7.61.
20 **N451T** CL-44D-4 *D.* FTL 1.6.61.
21 **N452T** CL-44D-4 *D.* FTL 23.6.61.
22 **N453T** CL-44D-4 *D.* FTL.
23 **N125SW** CL-44D-4 *D.* Seaboard 7.7.61.
24 **N454T** CL-44D-4 *D.* FTL.
25 **N455T** CL-44D-4 *D.* FTL 20.7.61.
26 **N126SW** CL-44D-4 *D.* Seaboard 30.9.61.
27 **N127SW** CL-44D-4 *D.* Seaboard.
28 **N602SA** CL-44D-4 *D.* Slick 17.1.62.
29 **N603SA** CL-44D-4 *D.* Slick.
30 **N128SW** CL-44D-4 *D.* Seaboard.

CONVAIR 880

22-00-1 **N8489H** Model 22-1: *F.F.* 27.1.59 as N801TW. *Cvtd.* to 880-M. *F.F.* 3.10.60.
22-00-2 **N802TW** Model 22-1: Static tests. *D.* TWA.
22-00-3 **N803TW** Model 22-1: *F.F.* 31.3.59. *D.* TWA.

22-00-4 **N8801E** Model 22-2: *F.F.* 10.8.59. *D.* Delta.
22-00-5 **N8478H** Model 22-1: *D.* Northeast (L.) 21.1.61 (ex TWA N804TW).
22-00-6 **N805TW** Model 22-1: *D.* TWA.
22-00-7 **N8802E** Model 22-2: *D.* Delta 31.1.60.
22-00-8 **N8479H** Model 22-1: *D.* Northeast (L.) 30.1.61 (ex TWA N806TW).
22-00-9 **N807TW** Model 22-1: *D.* TWA.
22-00-10 **N808TW** Model 22-1: *D.* TWA 18.5.60.
22-00-11 **N8803E** Model 22-2: *D.* Delta 4.5.60.
22-00-12 **N8480H** Model 22-1: *D.* Northeast (L.) 14.2.61 (ex TWA N809TW).
22-00-13 **N810TW** Model 22-1: *D.* TWA 15.2.61.
22-00-14 **N811TW** Model 22-1: *D.* TWA 2.2.61.
22-00-15 **N812TW** Model 22-1: *D.* TWA.
22-00-16 **N8804E** Model 22-2: *D.* Delta 6.5.60. *W.O.* 23.5.60.
22-00-17 **N8805E** Model 22-2: *D.* Delta 2.6.60.
22-00-18 **N813TW** Model 22-1: *D.* TWA.
22-00-19 **N814TW** Model 22-1: *D.* TWA.
22-00-20 **N8481H** Model 22-1: *D.* Northeast (L.) 8.12.60 (ex TWA N815TW).
22-00-21 **N8806E** Model 22-2: *D.* Delta 4.7.60.
22-00-22 **N8482H** Model 22-1: *D.* Northeast (L.) 5.12.60 (ex TWA N816TW).
22-00-23 **N8483H** Model 22-1: *D.* Northeast (L.) 30.11.60 (ex TWA N817TW).
22-00-24 **N818TW** Model 22-1: *D.* TWA 5.1.61.
22-00-25 **N819TW** Model 22-1: *D.* TWA 12.1.61.
22-00-26 **N820TW** Model 22-1: *D.* TWA 20.3.61.
22-00-27 **N821TW** Model 22-1: *D.* TWA 8.1.61.
22-00-28 **N822TW** Model 22-1: *D.* TWA 6.1.61.
22-00-29 **N8807E** Model 22-2: *D.* Delta 4.8.60.
22-00-30 **N823TW** Model 22-1: *D.* TWA 15.3.61.
22-00-31 **N824TW** Model 22-1: *D.* TWA.
22-00-32 **N825TW** Model 22-1: *D.* TWA.
22-00-33 **N826TW** Model 22-1: *D.* TWA.
22-00-34 **N827TW** Model 22-1: *D.* TWA.
22-00-35 **N828TW** Model 22-1: *D.* Hughes Tool Co.
22-00-36 **N8808E** Model 22-2: *D.* Delta 10.10.60.
22-00-37 — Model 22-3: Convair.
22-00-38 **N8809E** Model 22-2: *D.* Delta (L.) 21.10.60.
22-00-39 **N829TW** Model 22-1: *D.* Hughes Tool Co.
22-00-40 **N830TW** Model 22-1: *D.* Hughes Tool Co.
22-00-41 **N8810E** Model 22-2: *D.* Delta 21.11.60.
22-00-42 **N831TW** Model 22-1: *D* Hughes Tool Co.
22-00-43 — Model 22-3: Convair.
22-00-44 **B-1008** Model 22-4: *D.* Civil Air Transport.
22-00-45 — Model 22-3: Convair.
22-00-46 — Model 22-3: Convair.
22-00-47 **VR-HFS** Model 22-3: *D.* Cathay Pacific 1.4.62.
22-00-48 — Model 22-3: Convair.
22-00-49 — Model 22-3: Convair.
22-00-50 **N8811E** Model 22-2: *D.* Delta 9.9.61.
22-00-51 **N8812E** Model 22-2: *D.* Delta 20.9.61.
22-00-52 **N8813E** Model 22-2: *D.* Delta.
22-7-1 **YV-C-VIA** Model 22-21: *D.* Viasa.
22-7-2 **N8477H** Model 22-21: *D.* Alaska Airlines 25.7.61.
22-7-3 **N-112** Model 22-21: *D.* FAA.
22-7-4 **YV-C-VIB** Model 22-21: *D.* Viasa.
22-7-5 **JA-8021** Model 22-22: *D.* Japan Airlines 10.7.61 *Sakura.*
22-7-6 **JA-8022** Model 22-22: *D.* Japan Airlines *Matsu.*
22-7-7 **JA-8023** Model 22-22: *D.* Japan Airlines *Kaede.*
22-7-8 **JA-8024** Model 22-22: *D.* Japan Airlines.
22-7-9 **JA-8025** Model 22-22: *D.* Japan Airlines.

CONVAIR 990

30-10-1 **N5601** *F.F.* 24.1.61. *F.F.* after mods. 20.4.61.

30-10-2　N5602　*F.F.* 30.3.61.
30-10-3　N5603　*D.* American Airlines.
30-10-4　N5604　*D.* American Airlines.
30-10-5　HB-ICE　*D.* Swissair 1.62.
30-10-6　LN-LMA　Cancelled (SAS).
30-10-7　HB-ICA　*D.* Swissair 15.1.62.
30-10-8　SE-DAY　*D.* SAS 11.4.62. *Adils Viking. L.* Thai International.
30-10-9　N5605　*D.* American Airlines 7.1.62.
30-10-10　N5606　*D.* American Airlines 12.1.62.
30-10-11　HB-ICB　*D.* Swissair 1.62.
30-10-12　HB-ICC　*D.* Swissair 1.62.
30-10-13　—　Cancelled.
30-10-14　SE-DAZ　*D.* SAS.
30-10-15　HB-ICD　*D.* Swissair 1962.
30-10-16　N5607　*D.* American Airlines 10.2.62.
30-10-17　OY-KVA　Cancelled (SAS).
30-10-18　N5608　*D.* American Airlines 5.3.62.
30-10-24　N5609　*D.* American Airlines 20.3.62.
30-10-25　N5610　*D.* American Airlines.
30-10-26　N5611　*D.* American Airlines.
30-10-27　N5612　*D.* American Airlines.
30-10-31　N5613　*D.* American Airlines.
30-10-32　N5614　*D.* American Airlines.
30-10-33　N5615　*D.* American Airlines.

DE HAVILLAND COMET

0601　G-ALVG　Proto. Srs.1: *F.F.* 27.7.49 as G-5-1. *Sc.* 1953.
0602　G-ALZK　2nd proto. Srs.1: *F.F.* 27.7.50 as G-5-2. *Sc.* March 1957.
0603　G-ALYP　Srs.1: *F.F.* 9.1.51. First BOAC service 2.5.52. *W.O.* 10.1.54.
0604　G-ALYR　Srs.1: *F.F.* 28.7.51. BOAC. *Sc.* July 1955.
0605　G-ALYS　Srs.1: *F.F.* 8.9.51. *D.* BOAC 4.2.52. *Sc.* 1955.
0606　G-ALYT　Srs.2X: *F.F.* 16.2.52. Avon engine development.
0607　G-ALYU　Srs.1: *F.F.* 31.12.51. BOAC service. *Sc.* 1954.
0608　G-ALYV　Srs.1: *F.F.* 9.4.52. BOAC service. *W.O.* 2.5.53.
0609　G-ALYW　Srs.1: BOAC service. *Sc.* 1955.
06010　G-ALYX　Srs.1: BOAC service. *Sc.* 1955.
06011　G-ALYY　Srs.1: BOAC service. *W.O.* 8.4.54.
06012　G-ALYZ　Srs.1: *D.* BOAC 30.9.52. *W.O.* 26.10.52.
06013　G-ANAV　Srs.1A: *F.F.* 11.8.52 as CF-CUM for CPA. *D.* BOAC 1953. *Sc.* 1955.
06014　CF-CUN　Srs.1A: *D.* CPA and *W.O.* 3.3.53.
06015　F-BGSA　Srs.1A: *D.* UAT 17.12.52. *Withdrawn* April 1954.
06016　F-BGSB　Srs.1A: *D.* UAT. *Withdrawn* April 1954.
06017　5301　Srs.1A: *D.* RCAF No. 412 Sqdn. 29.5.53. *Cvtd.* to 1XB, *D.* 26.9.57.
06018　5302　Srs.1A: *D.* RCAF No. 412 Sqdn. 16.6.53. *Cvtd.* to 1XB. *D.* 26.9.57.
06019　F-BGSC　Srs.1A: *D.* UAT. *W.O.* 25.6.53.
06020　G-AOJT　Srs.1A: *D.* Air France 1953. as F-BGNX *D.* Min. of S.
06021　XM829　Srs.1A: *D.* Air France 1953. as F-BGNY. To Min. of S. as G-AOJU. *Cvtd.* to 1XB.
06022　XM823　Srs.1A: *D.* Air France 1953. as F-BGNZ. To Min. of S. as G-APAS. *Cvtd.* to 1XB.
06023　XK655　Srs.2: *F.F.* 27.8.53 as G-AMXA. To RAF for radar trials.
06024　XK669　Srs.2: *F.F.* 3.11.53 as G-AMXB. To RAF as T.Mk.2 for 216 Sqdn. *Taurus.*
06025　XK659　Srs.2: *F.F.* as G-AMXC. To RAF for radar trials.
06026　XN453　Srs.2: *F.F.* as G-AMXO. *Cvtd.* to 2E and *F.F.* 4.57 for Avon development. To RAE for radio trials.
06027　XK663　Srs.2: *F.F.* as G-AMXE. To RAF for radar trials.
06028　XK670　Srs.2: *F.F.* as G-AMXF. To RAF as T.Mk.2 for 216 Sqdn. *Corvus.*
06029　XK671　Srs.2: Was G-AMXG. To RAF as C.Mk.2 for 216 Sqdn. *Aquila.*
06030　XK695　Srs.2: Was G-AMXH. To RAF as C.Mk.2 for 216 Sqdn. *Perseus.*
06031　XK696　Srs.2: Was G-AMXI. To RAF as C.Mk.2 for 216 Sqdn. *Orion.*

06032　XK697　Srs.2: Was G-AMXJ. To RAF as C.Mk.2 for 216 Sqdn. *Cygnus.*
06033　G-AMXK　Srs.2: *Cvtd.* to 2E for Avon 29 development.
06034　XK698　Srs.2: Was G-AMXL. To RAF as C.Mk.1 for 216 Sqdn. *Pegasus.*
06035　XK699　Srs.2: To RAF as C.Mk.1 for 216 Sqdn. *Sagittarius.*
06036　—　Srs.2: not flown. Structural tests.
06037　XK715　Srs.2: To RAF as C.Mk.1 for 216 Sqdn. *Columba.*
06045　XK716　Srs.2: To RAF as C.Mk.1 for 216 Sqdn. *Cepheus.*
06100　XP915　Srs.3: *F.F.* 19.7.54 as G-ANLO. *Cvtd.* to 3B, *F.F.* 21.8.58. To BLEU 1961.
6401　G-APDA　Srs.4: *F.F.* 27.4.58. *D.* BOAC 24.2.59.
6402　—　Srs.4: for water tank tests.
6403　G-APDB　Srs.4: *F.F.* 27.7.58. *D.* BOAC 30.9.58.
6404　G-APDC　Srs.4: *F.F.* 23.9.58. *D.* BOAC 30.9.58.
6405　G-APDD　Srs.4: *F.F.* 5.11.58. *D.* BOAC 18.11.58.
6406　G-APDE　Srs.4: *F.F.* 20.9.58. *D.* BOAC 2.10.58.
6407　G-APDF　Srs.4: *F.F.* 11.12.58. *D.* BOAC 31.12.58.
6408　LV-AHN　Srs.4: *F.F.* 27.1.59. *D.* Aerolineas 2.3.59 as LV-PLM. *Los Tres Marias.*
6409　G-APDH　Srs.4: *F.F.* 21.11.59. *D.* BOAC 6.12.59.
6410　LV-AHO　Srs.4: *F.F.* 25.2.59. *D.* Aerolineas 18.3.59 as LV-PLN. *W.O.* 20.2.60.
6411　LV-AHP　Srs.4: *F.F.* 24.3.59. *D.* Aerolineas 2.5.59 as LV-PLP. *W.O.* 26.8.59.
6412　G-APDK　Srs.4: *F.F.* 2.1.59. *D.* BOAC 12.2.59.
6413　G-APDL　Srs.4: *F.F.* 27.4.59. *D.* BOAC 6.5.59.
6414　G-APDM　Srs.4: *F.F.* 21.3.59. *D.* BOAC 16.4.59.
6415　G-APDN　Srs.4: *F.F.* 29.5.59. *D.* BOAC 10.6.59.
6416　G-APDO　Srs.4: *F.F.* 29.4.59. *D.* BOAC 14.5.59.
6417　G-APDP　Srs.4: *F.F.* 29.5.59. *D.* BOAC 11.6.59.
6418　G-APDR　Srs.4: *F.F.* 9.7.59. *D.* BOAC 20.7.59.
6419　G-APDS　Srs.4: *F.F.* 6.8.59. *D.* BOAC 14.8.59.
6420　G-APDT　Srs.4: *F.F.* 2.10.59. *D.* BOAC 19.10.59.
6421　G-APMA　Srs.4B: *F.F.* 27.6.59. *D.* BEA 20.12.59. *Edmond Halley.*
6422　G-APMB　Srs.4B: *F.F.* 29.7.59. *D.* BEA 9.11.59 *Walter Gale.*
6423　G-APMC　Srs.4B: *F.F.* 1.10.59. *D.* BEA 16.11.59. *Andrew Grommelin.*
6424　XA-NAR　Srs.4C: *F.F.* 31.10.59 as G-AOVU. *D.* Mexicana 1960.
6425　XA-NAS　Srs.4C: *F.F.* 3.12.59 as G-AOVV. *D.* Mexicana 13.12.59.
6426　G-APMF　Srs.4B: *F.F.* 5.1.60. *D.* BEA 27.1.60. *William Finlay.*
6427　G-APDG　Srs.4: *F.F.* 12.11.59. *D.* BOAC 7.12.59.
6428　G-APDI　Srs.4: *F.F.* 7.12.59. *D.* BOAC 18.12.59.
6429　G-APDJ　Srs.4: *F.F.* 23.12.59. *D.* BOAC 11.1.60.
6430　LV-AHR　Srs.4: *F.F.* 15.2.60. *D.* Aerolineas 8.3.60 as LV-FOY. *Arco Iris. W.O.* 23.11.61.
6431　VP-KPJ　Srs.4: *F.F.* 14.7.60. *D.* EAAC. 25.7.60.
6432　LV-AHS　Srs.4: *F.F.* 18.2.60. *D.* Aerolineas 19.3.60 as LV-POZ. *Alborada.*
6433　VP-KPK　Srs.4: *F.F.* 28.7.60. *D.* EAAC 6.9.60.
6434　LV-AHU　Srs.4: *F.F.* 2.7.60. *D.* Aerolineas 26.7.60 as LV-PPA. *Centaurus.*
6435　G-APMD　Srs.4B: *F.F.* 16.3.60. *D.* BEA 29.3.60. *William Denning.*
6436　G-APME　Srs.4B: *F.F.* 26.4.60. *D.* BEA 10.5.60. *John Tebbutt.*
6437　SX-DAK　Srs.4B: *F.F.* 8.4.60 as G-APYC. *D.* Olympic 26.4.60. *Queen Frederica.*
6438　SX-DAL　Srs.4B: *F.F.* 3.5.60 as G-APYD. *D.* Olympic 14.5.60. *Queen Olga.*
6439　SU-ALC　Srs.4C: *F.F.* 21.5.60. *D.* UAA 10.6.60.
6440　G-APZM　Srs.4B: *F.F.* 30.6.60. *D.* Olympic 14.7.60. *Queen Sophia.*
6441　SU-ALD　Srs.4C: *F.F.* 15.6.60. *D.* UAA 29.6.60.
6442　G-APMG　Srs.4B: *F.F.* 25.7.60. *D.* BEA 31.7.60.
6443　XA-NAT　Srs.4C: *F.F.* 7.10.60 as G-ARBB. *D.* Mexicana 29.11.60.
6444　SU-ALE　Srs.4C: *F.F.* 22.11.60. *D.* UAA 23.12.60.
6445　OD-ADR　Srs.4C: *F.F.* 3.12.60. *D.* MEA 19.12.60.
6446　OD-ADQ　Srs.4C: *F.F.* 7.2.61 as G-ARJH. *D.* MEA 15.2.61.
6447　G-ARDI　Srs.4B: *F.F.* 24.3.61. *D.* Olympic 25.3.61. *Princess Sophia.*
6448　OD-ADS　Srs.4C: *F.F.* 4.3.61. *D.* MEA 14.3.61.
6449　G-ARCO　Srs.4B: *F.F.* 5.4.61. *D.* BEA 13.4.61.
6450　OD-ADT　Srs.4C: *F.F.* 9.3.61. *D.* MEA 18.3.61.
6451　G-ARCP　Srs.4B: *F.F.* 12.4.61. *D.* BEA 19.4.61.

DE HAVILLAND COMET—*continued*

6452 G-ARJK   Srs.4B: *F.F.* 4.5.61. *D.* BEA 15.5.61.
6453 G-ARGM   Srs.4B: *F.F.* 27.4.61. *D.* 6.5.61.
6454 SU-ALL   Srs.4C: *F.F.* 30.5.61. *D.* UAA 12.6.61.
6455 G-ARJL   Srs.4B: *F.F.* 19.5.61. *D.* BEA 31.5.61.
6456 G-ARJM   Srs.4B: *F.F.* 8.6.61. *D.* BEA 26.6.61. *W.O.* 21.12.61.
6457 —
6458 SU-ALM   Srs.4C: *F.F.* 30.6.61. *D.* UAA 15.7.61.
6459 G-ARJN   Srs.4B: *F.F.* 21.7.61. *D.* BEA 4.8.61.
6460 LV-PTS   Srs.4C: *F.F.* 24.8.61 as G-AROV. *D.* Aerolineas Argentinas 26.4.62.
6461 SA.R.7   Srs.4C: King Ibn Saud.
6462 SU-AMV   Srs.4C: *D.* UAA 6.4.62.
6463 —
6464 SU-AMW   Srs.4C: UAA.
6465 —
6466 —
6467 XR395   C.Mk.4C: *F.F.* 15.11.61. No. 216 Sqdn. RAF.
6468 XR396   C.Mk.4C: No. 216 Sqdn. RAF.
6469 XR397   C.Mk.4C: *F.F.* 18.1.62. No. 216 Sqdn. RAF.
6470 XR398   C.Mk.4C: *F.F.* 13.2.62. No. 216 Sqdn. RAF.
6471 XR399   C.Mk.4C: No. 216 Sqdn. RAF.
6472 VP-KRL   Srs.4: *D.* EAAC 10.4.62.

DOUGLAS DC-8

45252 N8008D   *F.F.* 30.5.58. *Cvtd.* Srs.50. *F.F.* 20.12.60. *D.* National 20.10.61.
45253 N800PA   Srs.30: *D.* PAA 2.6.61. *Flying Cloud.*
45254 N801PA   Srs.30: *D.* PAA 8.4.61. *Queen of the Pacific.*
45255 N802PA   Srs.30: *F.F.* as N8068D *D.* PAA 16.2.61. *Cathay.*
45256 N803PA   Srs.30: *D.* PAA 7.2.60. *Mandarin.*
45257 N804PA   Srs.30: *D.* PAA 17.3.60. *Midnight Sun.*
45258 N805PA   Srs.30: *D.* PAA 20.3.60. *Nightingale.*
45259 N806PA   Srs.30: *D.* PAA 7.5.60. *Northern Light.*
45260 N807PA   Srs.30: *D.* PAA 3.6.60. *Polynesia.*
45261 N808PA   Srs.30: *D.* PAA 11.6.60. *Gauntlet.*
45262 N809PA   Srs.30: *D.* PAA 23.6.60. *Great Republic.*
45263 N810PA   Srs.30: *D.* PAA 17.7.60. *Intrepid.*
45264 N811PA   Srs.30: *D.* PAA 22.8.60. *Pacific Trader.*
45265 N812PA   Srs.30: *D.* PAA 10.9.60. *Blue Jacket.*
45266 N813PA   Srs.30: *D.* PAA 14.10.60. *Bostonian.*
45267 N814PA   Srs.30: *D.* PAA 19.10.60. *Caroline.*
45268 N815PA   Srs.30: *D.* PAA 8.11.60. *Charger.*
45269 N816PA   Srs.30: *D.* PAA 11.11.60. *East Indian.*
45270 N817PA   Srs.30: *D.* PAA 16.12.60. *Derby.*
45271 N818PA   Srs.30: *D.* PAA 23.12.60. *Rambler.*
45272 PP-PDS   Srs.30: *D.* Panair do Brasil 27.2.61. *Manoel de Borba Gato.*
45273 PP-PDT   Srs.30: *D.* Panair do Brasil 27.2.61. *Bras Cubas.*
45274 N8274H   Srs.30: *D.* Panagra 7.4.60.
45275 N8275H   Srs.30: *D.* Panagra 25.5.60.
45276 N8276H   Srs.30: *D.* Panagra 29.6.60.
45277 N8277H   Srs.30: *D.* Panagra 30.9.60.
45278 N8001U   Srs.10: *F.F.* 29.11.58 as N8018D. *D.* United 1.11.60.
45279 N8002U   Srs.10: *F.F.* as N8028D. *D.* United 11.5.61.
45280 N8003U   Srs.10: *F.F.* as N8038D. *D.* United 16.6.60. *Capt. Ralph J. Johnson.*
45281 N8004U   Srs.10: *D.* United 29.5.59. *Capt. R. T. Freng.*
45282 N8005U   Srs.10: *D.* United 27.6.59. *Capt. W. D. Williams.*
45283 N8006U   Srs.10: *D.* United 20.8.59. *Capt. R. L. Dobie.*
45284 N8007U   Srs.10: *D.* United 25.8.59. *Harry Hucking.*
45285 N8008U   Srs.10: *D.* United 3.9.59. *Capt. R. L. Wagner.*
45286 N8009U   Srs.10: *D.* United 14.9.59. *Capt. George Douglass.*
45287 N8010U   Srs.10: *D.* United 29.9.59. *Donald Douglas, Sr.*
45288 N8011U   Srs.10: *D.* United 22.10.59. *Wm. B. Stout.*

45289 N8012U   Srs.10: *D.* United 18.12.59. *J. A. Herlihy.*
45290 N8013U   Srs.10: *D.* United 22.12.59. *Will Rogers. W.O.* 16.12.60.
45291 N8018U   Srs.10: *D.* United 28.2.60. *Hana-Maui.*
45292 N8023U   Srs.20: *D.* United 26.2.60. *Waipahu.*
45293 N8024U   Srs.20: *D.* United 6.4.60. *Hilo.*
45294 N8025U   Srs.20: *D.* United 12.4.60. *Annie E. Johnson.*
45295 N8026U   Srs.20: *D.* United 17.5.60. *Oahu.*
45296 N8027U   Srs.20: *D.* United 25.5.60.
45297 N8028U   Srs.20: *D.* United 8.6.60.
45298 N8029U   Srs.20: *D.* United 17.8.60.
45299 N8031U   Srs.20: *D.* United 11.9.60.
45300 N8033U   Srs.20: *D.* United 13.10.60.
45301 N8034U   Srs.50: *D.* United 2.5.61.
45302 N8035U   Srs.50: *D.* United 18.5.61.
45303 N8036U   Srs.50: *D.* United 8.6.61.
45304 N8037U   Srs.10: *D.* United 25.1.61.
45305 N8038U   Srs.10: *D.* United 8.2.61.
45306 N8039U   Srs.10: *D.* United 11.4.61.
45307 N8040U   Srs.10: *D.* United 16.6.61. *W.O.* 11.7.61.
45376 PH-DCA   Srs.30: *D.* KLM 19.3.60. *Albert Plesman.*
45377 PH-DCB   Srs.30: *D.* KLM 19.4.60. *Daniel Bernoulli.*
45378 PH-DCC   Srs.30: *D.* KLM 9.6.60. *Sir Frank Whittle.*
45379 PH-DCD   Srs.30: *D.* KLM 5.7.60. *Nikolaus Otto.*
45380 PH-DCE   Srs.30: *D.* KLM 20.8.60. *Thomas Alva Edison.*
45381 PH-DCF   Srs.30: *D.* KLM 25.9.60. *Anthony Fokker.*
45382 PH-DCG   Srs.30: *D.* KLM 27.10.60. *Guglielmo Marconi.*
45383 PH-DCH   Srs.50: *F.F.* as N96032. *D.* KLM 14.7.61. *Orville Wright.*
45384 OY-KTA   Srs.30: *D.* SAS 31.3.60. *Dan.*
45385 LN-MOA   Srs.30: *D.* SAS 4.5.60. *Haakon.*
45386 SE-DBA   Srs.30: *D.* SAS 11.5.60. *Rurik.*
45387 OY-KTB   Srs.30: *D.* SAS 14.7.60. *Blue.*
45388 LN-MOT   Srs.30: *D.* SAS 1.9.60. *Olav.*
45389 SE-DBB   Srs.30: *D.* SAS 2.10.60. *Ottar.*
45390 SE-DBC   Srs.30: *D.* SAS 11.11.60. *Visbur.*
45391 N6571C   Srs.20: *D.* National 7.2.60.
45392 N6572C   Srs.20: *D.* National 17.3.60.
45393 N6573C   Srs.20: *D.* National 9.12.60.
45408 N801E   Srs.10: *D.* Delta 23.7.59. *Westward Ho.*
45409 N802E   Srs.10: *D.* Delta 14.9.59.
45410 N803E   Srs.10: *D.* Delta 10.10.59.
45411 N804E   Srs.10: *D.* Delta 28.10.59.
45412 N805E   Srs.10: *D.* Delta 4.11.59.
45413 N806E   Srs.10: *D.* Delta 12.11.59.
45416 HB-IDA   Srs.30: *D.* Swissair 23.4.60. *Geneve.*
45417 HB-IDB   Srs.30: *D.* Swissair 19.6.60. *Basel Stadt.*
45418 JA8001   Srs.30: *D.* JAL 17.7.60. *Fuji.*
45419 JA8002   Srs.30: *D.* JAL 29.7.60. *Nikko.*
45420 JA8003   Srs.30: *D.* JAL 13.9.60. *Hakkone. W.O.* 24.4.61.
45421 JA8005   Srs.30: *D.* JAL 24.11.60. *Miyajima.*
45422 N8601   Srs.20: *D.* Eastern 3.1.60.
45423 N8602   Srs.20: *D.* Eastern 22.1.60.
45424 N8603   Srs.20: *D.* Eastern 14.2.60.
45425 N8604   Srs.20: *D.* Eastern 19.2.60.
45426 N8605   Srs.20: *D.* Eastern 8.4.60.
45427 N8606   Srs.20: *D.* Eastern 26.4.60.
45428 N8607   Srs.20: *D.* Eastern 22.5.60.
45429 N8608   Srs.20: *D.* Eastern 8.8.60.
45430 N8609   Srs.20: *D.* Eastern 9.10.60.
45431 N8610   Srs.20: *D.* Eastern 23.10.60.
45432 XA-XAX   Srs.20: *D.* Aeronaves de Mexico 28.10.60. *20 de Noviembre.*
          W.O. 19.1.61.

45433	N8612	Srs.20: _D._ Eastern 15.11.60.
45434	N8613	Srs.20: _D._ Eastern 28.7.61.
45435	N8614	Srs.20: _D._ Eastern 15.8.61.
45436	N8615	Srs.20: _D._ Eastern 6.9.61.
45437	N8617	Srs.20: _D._ Eastern 23.10.61.
45442	CF-TJA	Srs.40: _F.F._ as N6577C. _D._ TCA 25.9.60.
45443	CF-TJB	Srs.40: _F.F._ as N6578C. _D._ TCA 25.5.60.
45444	CF-TJC	Srs.40: _D._ TCA 27.3.60.
45445	CF-TJD	Srs.40: _D._ TCA 7.2.60.
45526	HB-IDC	Srs.30: _D._ Swissair 30.8.60. _Zurich._
45565	CF-TJE	Srs.40: _D._ TCA 18.11.60.
45566	CF-TJF	Srs.40: _D._ TCA 17.12.60.
45567	F-BJLA	Srs.30: _F.F._ as N9601Z. _D._ UAT 28.6.60.
45568	F-BJLB	Srs.30: _D._ UAT 5.8.60.
45569	F-BIUY	Srs.30: _D._ TAI 31.7.60.
45570	F-BIUZ	Srs.30: _D._ TAI 1.3.61.
45588	N8014U	Srs.10: _D._ United 18.11.59.
45589	N8015U	Srs.10: _D._ United 22.11.59. _W. E. Rhoades._
45590	N8016U	Srs.10: _D._ United 2.12.59. _W. C. Mentzer._
45591	N8017U	Srs.10: _D._ United 5.12.59. _James Doolittle._
45592	N8019U	Srs.10: _D._ United 30.12.59.
45593	N8020U	Srs.10: _D._ United 14.1.60.
45594	N8021U	Srs.20: _D._ United 21.1.60.
45595	N8022U	Srs.20: _D._ United 17.2.60.
45596	N8030U	Srs.20: _D._ United 27.8.60.
45597	N8032U	Srs.20: _D._ United 29.9.60.
45598	I-DIWA	Srs.40: _D._ Alitalia 28.4.60. _Amerigo Vespucci._
45599	I-DIWE	Srs.40: _D._ Alitalia 5.7.60. _Christoforo Colomba._
45600	I-DIWI	Srs.40: _D._ Alitalia 22.7.60. _Giovanni da Verazzano._
45601	I-DIWO	Srs.40: _D._ Alitalia 3.11.60. _Marco Polo._
45602	N801US	Srs.30: _D._ Northwest 17.5.60.
45603	N802US	Srs.30: _D._ Northwest 3.7.60.
45604	N803US	Srs.30: _D._ Northwest 11.8.60.
45605	N804US	Srs.30: _D._ Northwest 22.9.60.
45606	N805US	Srs.30: _D._ Northwest 5.1.61.
45607	N9608Z	Srs.50: Philippine Airlines. _Pacific Pacer._
45608	—	Srs.50: Philippine Airlines.
45609	CF-TJG	Srs.40: _D._ TCA 22.12.60.
45610	CF-TJH	Srs.40: _D._ TCA 13.1.61.
45611	CF-TJI	Srs.40: _D._ TCA 2.2.61.
45612	CF-TJJ	Srs.40: _D._ TCA 23.3.61.
45613	PH-DCI	Srs.50: _D._ KLM 4.4.61. _Sir Isaac Newton._
45614	PH-DCK	Srs.50: _F.F._ as N96052. _D._ KLM 2.5.61. _Admiral Richard E. Byrd._
45615	PH-DCL	Srs.50: _D._ KLM 2.5.61. _W.O._ 30.5.61.
45616	PH-DCM	Srs.50: _D._ KLM 25.6.61.
45617	EC-ARA	Srs.50: _D._ Iberia 22.5.61. _Velazquez._
45618	EC-ARB	Srs.50: _D._ Iberia 27.5.61. _El Greco._
45619	EC-ARC	Srs.50: _D._ Iberia 20.6.61. _Goya._
45620	CF-CFF	Srs.40: _D._ CPA 22.2.61. _Vancouver._
45621	CF-CPG	Srs.40: _F.F._ as N96042. _D._ CPA 1.4.61.
45622	CF-CPH	Srs.40: _D._ CPA 20.5.61. _Calgary._
45623	CF-CPI	Srs.40: _D._ CPA 75.11.61.
45624	I-DIWB	Srs.40: _D._ Alitalia 29.4.61.
45625	I-DIWU	Srs.40: _D._ Alitalia 31.5.61.
45626	JA8006	Srs.30: _D._ JAL 5.5.61. _Kamakura._
45627	F-BJUV	Srs.30: _D._ TAI 3.6.61.
45628	N8780R	Srs.50: _D._ Trans Caribbean 6.11.61. _James Roy._
45629	PH-DCN	Srs.50: _D._ KLM 18.1.62. _Albert Schweitzer._
45630	I-DIWF	Srs.40: _D._ Alitalia 28.2.62.
45631	I-	Srs.40: _D._ Alitalia.
45632	N-	

45633	XA-	Srs.50: _D._ Aeronaves. 5.62. _Moctezuma-Ilhuicamina._
45634	N-774C	Srs.50: _D._ National 23.2.62.
45635	N-	
45636	I-DIWP	Srs.40: _D._ Alitalia 3.11.61.
45637	I-DIWR	Srs.40: _D._ Alitalia 1.2.62.
45638	CF-TJK	Srs.40: _D._ TCA 16.12.61.
	JA-8007	Srs.50: _D._ JAL 3.62. _Yoshino._

FAIRCHILD F-27

1	HC-ADV	_F.F._ as N1027, 12.4.58. _Cvtd._ to F-27A. _D._ AREA 26.6.59. _W.O._ 7.11.60.
2	XA-MOT	_F.F._ as N2027, 23.5.58. _D._ Trans-Mar de Cortes 10.4.59.
3	N2701	F-27: _D._ West Coast 24.7.58.
4	N2700R	F-27: _D._ Piedmont 6.10.58.
5	N2702	F-27: _D._ West Coast 11.8.58.
6	N2703	F-27: _D._ West Coast 8.8.58.
7	N2704	F-27: _D._ West Coast 14.8.58.
8	N2701R	F-27: _D._ Piedmont 1.10.58.
9	N2702R	F-27: _D._ Piedmont 6.10.58.
10	N2703R	F-27: _D._ Piedmont 20.10.58.
11	CF-QBA	F-27: _D._ Quebecair 17.9.58.
12	YV-C-EVH	F-27: _D._ Avensa 18.9.58. _W.O._ 25.2.62.
13	YV-C-EVK	F-27: _D._ Avensa 25.9.58.
14	CF-QBZ	F-27: _D._ Quebecair 6.10.58.
15	N4903	F-27B: _D._ Northern Consolidated 28.10.58.
16	N2705	F-27: _D._ West Coast 23.10.58.
17	N2704R	F-27: _D._ Piedmont 31.10.58.
18	N2705R	F-27: _D._ Piedmont 31.10.58.
19	N2706R	F-27: _D._ Piedmont 26.11.58.
20	N2707R	F-27: _D._ Piedmont 26.11.58.
21	N4904	F-27B: _D._ Northern Consolidated 26.11.58.
22	N2706	F-27: _D._ West Coast 12.12.58.
23	N5G	F-27: _D._ General Tire and Rubber 29.11.58.
24	YV-C-EVS	F-27: _D._ Avensa 5.12.58.
25	YV-C-EVP	F-27: _D._ Avensa 3.12.58.
26	YV-C-EVQ	F-27: _D._ Avensa 5.12.58.
27	N994	F-27: _D._ Reynolds Metal 30.12.58.
28	N1500	F-27A: _D._ Continental Carr. 13.1.59.
29	N145L	F-27A: _D._ Bonanza 16.1.59.
30	XC-CAT	F-27: _D._ Bank of Mexico 17.1.59.
31	N1924	F-27A: _D._ Ideal Cement 13.1.59.
32	N3027	F-27: _D._ H. K. Porter Inc. 8.5.59.
33	N1004	F-27A: _D._ Butler 10.4.59.
34	N65A	F-27: _D._ Noland Co. 4.2.59.
35	N100L	F-27A: _D._ Westinghouse 5.2.59.
36	N2770R	F-27A: _D._ Pacific Airlines 19.2.59.
37	N146L	F-27A: _D._ Bonanza 17.2.59.
38	N147L	F-27A: _D._ Bonanza 26.2.59.
39	N2708R	F-27A: _D._ Wien Alaska 22.5.59.
40	N5095A	F-27: _D._ Aloha 13.5.59.
41	N200KC	F-27A: _D._ Kimberley-Clark Corp. 13.3.59.
42	N1823A	F-27A: _D._ Champion Spark Plug 25.3.59.
43	N2771R	F-27A: _D._ Pacific Airlines 2.4.59.
44	N2772R	F-27A: _D._ Pacific Airlines 18.4.59.
45	N25JM	F-27: _D._ Johns Manville Corp. 23.4.59.
46	N5093A	F-27: _D._ Aloha 14.4.59.
47	CF-QBL	F-27: _D._ Quelecair 27.7.59.
48	N2712	F-27A: _D._ Bonanza 2.6.61.
49	N4905	F-27B: _D._ Northern Consolidated 30.4.59.
50	N5094A	F-27: _D._ Aloha 8.5.59.
51	N2773R	F-27A: _D._ Pacific Airlines 28.5.59.
52	N2774R	F-27A: _D._ Pacific Airlines 11.6.59.

53	N2775R	F-27A: *D.* Pacific Airlines 24.6.59.
54	N200R	F-27A: *D.* Raytheon 22.6.59.
55	N2710R	F-27A: *D.* Wien Alaska 30.6.59.
56	N42N	F-27A: *D.* Union Carbide 21.7.59.
57	N27C	F-27A: *D.* Murray Corp. 20.1.60.
58	N4300F	F-27: *D.* Ozark 16.7.59.
59	N4301F	F-27: *D.* Ozark 13.8.59.
60	N4302F	F-27: *D.* Ozark 12.8.59.
61	N27R	F-27: *D.* Reynolds Tobacco Co. 13.8.59.
62	N148L	F-27A: *D.* Bonanza 17.8.59.
63	N149L	F-27A: *D.* Bonanza 17.9.59.
64	N270VR	F-27A: *D.* Whitney 3.10.59.
65	N150L	F-27A: *D.* Bonanza 17.8.59.
66	N42Q	F-27A: *D.* Bonanza 17.9.59.
67	N12500	F-27A: *D.* U.S. Steel 12.11.59.
68	N991	F-27A: *D.* Reynolds Metal 1.12.59.
69	N1027	F-27A: *D.* Landis Tool 7.1.61.
70	N1X	F-27A: *D.* Olin Mathieson 8.4.60.
71	N1100L	F-27A: *D.* Westinghouse Electric 30.1.60.
72	CF-LWN	F-27: *D.* Abiti Power and Paper 4.2.60.
73	N5096A	F-27: *D.* Aloha Airline 16.2.60.
74	N42Q	F-27A: *D.* Union Carbide 7.3.60.
75	N5097A	F-27: *D.* Aloha Airline 14.3.60.
76	—	F-27A: *D.* Olin Mathieson 16.6.61.
77	N151L	F-27A: *D.* Bonanza 7.5.60.
78	N152L	F-27A: *D.* Bonanza 14.5.60.
79	—	F-27A: *D.* Gates Rubber 9.3.61.
80	N5098A	F-27: *D.* Aloha Airline 25.5.60.
81	TC-KOC	F-27: *D.* THY 10.8.60.
82	TC-KOD	F-27: *D.* THY 10.8.60.
83	TC-KOP	F-27: *D.* THY 10.8.60. *W.O.* 8.3.62.
84	—	F-27A: *D.* IBM 29.3.61.
85	—	F-27A: *D.* Cameron Iron Works 12.4.61.
86	TC-KOR	F-27: *D.* THY 23.7.61.
87	TC-KOZ	F-27: *D.* THY 23.7.61.
88	—	F-27A:
89	—	
90	—	F-27F:
91	N153L	F-27F: *D.* Bonanza 1962.
92	N154L	F-27F: *D.* Bonanza 1962.

FOKKER F-27

10101	PH-NIV	*F.F.* 24.11.55.
10102	Static test airframe.	
10103	PH-NVF	*F.F.* 31.1.57. *L.* LTU 1961. *L.* Luxair 1962.
10104	Fatique test airframe.	
10105	EI-AKA	*F.F.* 23.3.58. *D.* Aer Lingus 19.11.58. *Fionntan.*
10106	EI-AKB	*D.* Aer Lingus 19.11.58. *Fearghal.*
10107	EI-AKC	*D.* Aer Lingus 11.12.58. *Fionnbharr.*
10108	LN-SUN	*F.F.* 30.10.58. *D.* Braathens 20.12.58.
10109	EI-AKD	*F.F.* 24.12.58. *D.* Aer Lingus 23.1.59. *Flannan.*
10110	EI-AKE	*F.F.* 14.1.59. *D.* Aer Lingus 4.2.59. *Feidhlim.*
10111	VH-TFA	*F.F.* 16.1.59. *D.* TAA 13.6.59.
10112	VH-TFB	*D.* TAA 6.4.59. *W.O.* 10.6.60.
10113	VH-TFC	*D.* TAA 18.6.59.
10114	VH-TFD	*D.* TAA 5.5.59.
10115	59-0259	*D.* Philippine Air Force 2.9.59.
10116	LN-SUO	*D.* Braathens 24.4.59.
10118	EI-AKF	*D.* Aer Lingus 6.5.59. *Finghin.*
10119	EI-AKG	*D.* Aer Lingus 12.5.59. *Fiachra.*

10120	VH-TFE	*D.* TAA 6.6.59.
10121	VH-TFF	*D.* TAA 24.6.59.
10122	VH-TFG	*D.* TAA 13.7.59.
10123	TC-TEZ	*F.F.* 16.8.60. *D.* THY 27.10.60.
10124	TC-TON	*F.F.* 16.9.60. *D.* THY 27.10.60.
10125	TC-TOY	*F.F.* 5.10.60. *D.* THY 27.10.60.
10126	EP-MRP	*F.F.* 28.8.59. *D.* Shah of Iran 12.9.59.
10127	VH-EWA	*F.F.* 16.7. 59. Ordered TAA as VH-TFH. *D.* EWA 29.7.59. *City of Tamworth.*
10131	VH-CAV	*F.F.* 22.8.59. *D.* Australian DCA 19.9.59.
10132	VH-CAT	*F.F.* 11.9.59. *D.* Australian DCA 28.9.59.
10133	VH-FNA	Srs.200: *F.F.* 20.9.59. *D.* Ansett/ANA 5.10.59.
10134	VH-TFI	*F.F.* 24.9.59. *D.* 5.10.59.
10135	VH-TFJ	*F.F.* 6.10.59. *D.* TAA 21.10.59.
10136	VH-FNB	Srs.200: *F.F.* 15.10.59. *D.* Ansett/ANA 30.10.59.
10137	D-BATU	Srs.200: *F.F.* 19.12.59. *D.* Horton GmbH 30.12.59.
10138	VH-TFK	*F.F.* 4.11.59. *D.* TAA 30.11.59.
10139	VH-MMS	Srs.200: *F.F.* 16.11.59. Ordered TAA as VH-TFL. *D.* MMA 7.12.59. *Swan.*
10140	PH-IOK	*F.F.* 12.12.59. *D.* Iranian Oil Co. 11.1.60.
10141	PH-IOP	*F.F.* 30.12.59. *D.* Iranian Oil Co. 22.1.60.
10142	PH-PBE	*F.F.* 18.3.60. *D.* Neth. Royal Flight 13.5.60.
10143	VH-FNC	Srs.200: *F.F.* 29.12.59. *D.* Ansett/ANA 4.2.60.
10144	VH-FND	Srs.200: *F.F.* 6.1.60. *D.* Ansett/ANA 12.2.60.
10145	VH-FNE	Srs.200: *F.F.* 20.1.60. *D.* Ansett/ANA for Queensland Airlines 25.2.60.
10146	VH-FNF	Srs.200: *F.F.* 26.1.60. *D.* Ansett/ANA 10.3.60.
10147	PI-C501	*F.F.* 10.2.60. *D.* PAL 23.2.60.
10148	PI-C502	*F.F.* 19.2.60. *D.* PAL 9.3.60.
10149	C-2	*F.F.* 15.6.60. *D.* R.Neth.A.F. 6.7.60.
10150	C-3	*F.F.* 28.6.60. *D.* R.Neth.A.F. 6.8.60.
10151	PH-IOS	*F.F.* 2.5.60. *D.* KLM 11.5.60.
10152	C-1	*F.F.* 14.7.60. *D.* R.Neth.A.F. 23.8.60.
10153	PH-IOT	*F.F.* 23.5.60. *D.* KLM 31.5.60.
10154	C-4	*F.F.* 18.7.60. *D.* R.Neth.A.F. 6.10.60.
10155	C-5	*F.F.* 26.8.60. *D.* R.Neth.A.F. 30.9.60.
10156	C-6	*F.F.* 19.9.60. *D.* R.Neth.A.F. 7.12.60.
10157	C-7	*F.F.* 30.9.60. *D.* R.Neth.A.F. 29.12.60.
10158	C-8	*F.F.* 6.10.60. *D.* R.Neth.A.F. 28.12.60.
10159	C-9	*F.F.* 19.10.60. *D.* R.Neth.A.F. 4.1.61.
10160	C-10	*F.F.* 4.11.60. *D.* R.Neth.A.F. 17.1.61.
10161	C-11	*F.F.* 23.11.60. *D.* R.Neth.A.F. 2.2.61.
10162	C-12	*F.F.* 7.3.61. *D.* R.Neth.A.F. 28.4.61.
10163	AP-ALM	Srs.200: *F.F.* 9.12.60. *D.* PIA 3.1.61.
10164	AP-ALN	Srs.200: *F.F.* 22.12.60. *D.* PIA 19.1.61.
10165	AP-ALO	Srs.200: *F.F.* 10.1.61. *D.* PIA 6.2.61.
10166	ZK-BXA	*F.F.* 7.11.60. *D.* NZNAC 26.11.60. *Kuaka.*
10167	ZK-BXB	*F.F.* 18.1.61. *D.* NZNAC 9.2.61. *Kotuku.*
10168	ZK-BXC	*F.F.* 3.2.61. *D.* NZNAC 24.2.61. *Kowhera.*
10169	ZK-BXD	*F.F.* 14.2.61. *D.* NZNAC 3.3.61. *Koreke.*
10170	VH-FNG	Srs.200: *F.F.* 28.2.61. *D.* Ansett/ANA 17.3.61.
10171	VT-DMA	*F.F.* 3.3.61. *D.* IAC 5.4.61.
10172	VT-DMB	*F.F.* 15.3.61. *D.* IAC 7.4.61.
10173	VT-DMC	*F.F.* 23.3.61. *D.* IAC 27.4.61.
10174	VT-DMD	*F.F.* 6.4.61. *D.* IAC 28.4.61.
10175	VT-DME	*F.F.* 26.4.61. *D.* IAC 4.5.61.
10176	LN-SUG	*F.F.* 9.5.61. *D.* Braathens 18.5.61.
10177	JA-8301	Srs.200: *F.F.* 23.5.61. *D.* All Nippon 30.5.61.
10178	JA-8302	Srs.200: *F.F.* 1.6.61. *D.* All Nippon 14.6.61.
10179	JA-8303	Srs.200: *F.F.* 13.6.61. *D.* All Nippon 20.6.61.
10180	VH-FNH	Srs.200: *F.F.* 23.6.61. *D.* Ansett/ANA 11.7.61.
10181	VH-FNI	Srs.200: *F.F.* 6.7.61. *D.* Ansett/ANA 14.7.61.

10182   TC-TAY   *F.F.* 11.7.61. *D.* THY 30.8.61. *W.O.* 23.9.61.
10183   TC-TEK   *F.F.* 25.7.61. *D.* THY 30.8.61.
10184   ZX-BXE   *F.F.* 22.8.61. *D.* NZNAC 30.8.61.
10185   ZX-BXF   *F.F.* 4.9.61. *D.* NZNAC 12.9.61.
10186   D-BAKU   Srs.300: *F.F.* 18.9.61. *D.* LTU 16.11.61. *Jan Willem.*
10187   AP-ALW   Srs.200: *F.F.* 6.10.61. *D.* PIA 19.10.61.
10188   AP-ALX   Srs.200: *F.F.* 20.10.61. *D.* PIA 6.11.61.
10189   ZK-BXG   *F.F.* 3.11.61. *D.* NZNAC 17.11.61.
10190   ZK-BXH   *F.F.* 20.11.61. *D.* NZNAC 30.11.61.
10191   PI-C503   *F.F.* 8.12.61. *D.* PAL 14.12.61.
10192   ST-AAA   Srs.200: *F.F.* 2.1.62 *D.* Sudan Airways 23.1.62.
10193   ST-AAR   Srs.200: *F.F.* 25.1.62. *D.* Sudan Airways 2.2.62.
10194   ST-AAS   Srs.200: *F.F.* 13.2.62. *D.* Sudan Airways 20.2.52.
10195   JA-8605   Srs.200: *F.F.* 6.2.62. *D.* All Nippon 20.2.62.
10196   JA-8606   Srs.200: *F.F.* 20.2.62. *D.* All Nippon 5.3.62.
10197   JA-8607   Srs.200: *F.F.* 8.3.62. *D.* All Nippon 23.3.62.
10198   PH-LIP   *F.F.* 27.3.62. *D.* Philips 7.4.62.
10199   LN-SUW   *F.F.* 9.4.62. *D.* Braathens.

HANDLEY PAGE **HERALD**

147   G-AODE   *F.F.* as HPR.3 25.8.55. *Cvtd.* HPR.7. *F.F.* 11.3.58. *W.O.* 30.8.58.
148   G-ARTC   *F.F.* as G-AODF HPR.3 8.56. *Cvtd.* HPR.7. *F.F.* 17.12.58. *Cvtd.* Series 200. *F.F.* 8.4.61.
149   G-APWA   Srs.100: *F.F.* 30.10.59. *L.* Jersey Airlines.
150   G-APWB   Srs.101: *L.* Jersey Airlines 27.7.61. *D.* BEA 1952.
151   G-APWC   Srs.101: *D.* BEA 3.1.62.
152   G-APWD   Srs.101: *D.* BEA.
153   G-APWE   Srs.201: *F.F.* 13.12.61. *D.* Jersey Airlines 3.1.62.
154   G-APWF   Srs.201: *D.* Jersey Airlines.
155   G-APWG   Srs.201: *D.* Jersey Airlines.
156   G-APWH   Srs.201: *D.* Jersey Airlines.
157   G-APWI   Srs.201: *D.* Jersey Airlines.
158   G-APWJ   Srs.201: *D.* Jersey Airlines.
159   CF-NAC   Srs.202: *D.* Nordair 13.2.62.
160   CF-MCK   Srs.202: *D.* MCA.
161   CF-NAF   Srs.202: *D.* Nordair 13.3.62.
162   CF-MCM   Srs.202: *D.* MCA.

LOCKHEED **ELECTRA**

1001   N174PS   *F.F.* 6.12.57 as N1881. *D.* PSA 1962.
1002   VR-HFM   *F.F.* 13.2.58 as N1882. *D.* Cathay Pacific 29.6.59.
1003   N1883   Static Tests, Demonstrations, P3V Test Vehicle (*F.F.* 19.8.58).
1004   VR-HFO   *F.F.* 10.4.58 as N1884. World Tour as N7144C. *D.* Cathay Pacific 1.4.5
1005   N5501   *F.F.* 19.5.58. *D.* EAL 3.1.59.
1006   N1R   *D.* General Motors 10.7.58 as N5501V. *D.* L.A. Dodgers 1961.
1007   N5502   *D.* EAL 8.10.58.
1008   N5503   *D.* EAL 20.10.58.
1009   N5504   *D.* EAL 2.11.58.
1010   N5505   *D.* EAL 10.11.58.
1011   N5506   *D.* EAL 3.11.58.
1012   N5507   *D.* EAL 14.11.58.
1013   N5509   *D.* EAL 29.11.58.
1014   N5510   *D.* EAL 13.1.59.
1015   N6101A   *D.* American Airlines 27.11.58. *New York. W.O.* 3.2.59.
1016   N5511   *D.* EAL 11.1.59.
1017   N5512   *D.* EAL 25.11.58.
1018   N5513   *D.* EAL 16.12.58.
1019   N6102A   *D.* American Airlines 10.12.58. *Chicago.*
1020   N5514   *D.* EAL 13.2.59.
1021   N5515   *D.* EAL 2.2.59.

1022   N5516   *D.* EAL 19.1.59.
1023   N5517   *D.* EAL 18.1.59.
1024   N6103A   *D.* American Airlines 4.1.59. *Detroit.*
1025   N6104A   *D.* American Airlines 9.1.59. *Washington.*
1026   N5518   *D.* EAL 14.2.59.
1027   N6105A   *D.* American Airlines 23.1.59. *Boston.*
1028   N6106A   *D.* American Airlines 27.1.59. *Dallas.*
1029   N5519   *D.* EAL 26.2.59.
1030   N5520   *D.* EAL 28.2.59.
1031   N6107A   *D.* American Airlines 5.2.59. *Fort Worth.*
1032   N5521   *D.* EAL 12.3.59.
1033   N5522   *D.* EAL 3.2.59.
1034   N5523   *D.* Eastern 25.2.59.
1035   N500IK   *D.* National 1.4.59.
1036   N5524   *D.* Eastern 26.2.59.
1037   N6108A   *D.* American Airlines 16.2.59. *Buffalo.*
1038   N5525   *D.* Eastern 28.3.59.
1039   VH-RMA   *D.* Ansett-ANA 27.2.59.
1040   N970IC   *D.* Braniff 29.4.59.
1041   N6109A   *D.* American Airlines 27.2.59. *Toronto.*
1042   N5526   *D.* Eastern 8.4.59.
1043   N5527   *D.* Eastern 16.4.59.
1044   VH-RMC   *D.* Ansett-ANA.
1045   N5528   *D.* Eastern 27.4.59.
1046   N7135C   *D.* Western 20.5.59.
1047   VH-RMB   *D.* Ansett-ANA 1.4.59.
1048   N5529   *D.* Eastern 7.5.59.
1049   N6110A   *D.* American Airlines 1.4.59. *St. Louis.*
1050   N6111A   *D.* American Airlines 16.4.59. *Tulsa.*
1051   N6112A   *D.* American Airlines 22.4.59. *Philadelphia.*
1052   N9702C   *D.* Braniff 6.5.59.
1053   N5530   *D.* Eastern 20.5.59.
1054   N6113A   *D.* American Airlines 30.4.59. *Syracuse.*
1055   N5531   *D.* Eastern 26.5.59.
1056   N6114A   *D.* American Airlines 13.5.59. *Rochester.*
1057   N121US   L-188C: *D.* Northwest 19.7.59. *W.O.* 17.3.60.
1058   N6115A   *D.* American Airlines 18.5.59. *New York.*
1059   N5002K   *D.* National 25.5.59.
1060   N5532   *D.* Eastern, 2.6.59.
1061   VH-TLA   *D.* TAA 15.6.59.
1062   N5533   *D.* Eastern 8.6.59. *W.O.* 4.10.60.
1063   N6116A   *D.* American Airlines 4.6.59. *Cincinatti.*
1064   N5003K   *D.* National 12.6.59.
1065   N6117A   *D.* American Airlines 17.6.59. *Louisville.*
1066   N5534   *D.* Eastern 25.6.59.
1067   N9703C   *D.* Braniff 22.6.59.
1068   N5535   *D.* Eastern 9.7.59.
1069   VH-TLB   *D.* TAA 14.7.59.
1070   N7136C   *D.* Western 10.7.59.
1071   N5536   L-188C: *D.* Eastern 27.7.59.
1072   N6118A   *D.* American Airlines 25.7.59. *Hartford.*
1073   N6119A   *D.* American Airlines 1.8.59. *Cleveland.*
1074   N7137C   *D.* Western 29.7.59.
1075   N5537   L-188C: *D.* Eastern 13.8.59.
1076   N5004K   *D.* National 6.8.59.
1077   N122US   L-188C: *D.* Northwest 29.7.59.
1078   N5538   L-188C: *D.* Eastern 20.8.59.
1079   N5005K   *D.* National 17.8.59.
1080   N5539   L-188C: *D.* Eastern 28.8.59.
1081   N6120A   *D.* American Airlines 17.8.59. *Newark.*
1082   N123US   L-188C: *D.* Northwest 11.8.59.

LOCKHEED ELECTRA—continued

1083 N6121A   *D.* American Airlines 23.8.59. *Providence.*
1084 N5006K   *D.* National 30.8.59.
1085 N124US   L-188C: *D.* Northwest 30.8.59.
1086 N9704C   *D.* Braniff 28.8.59.
1087 N7138C   *D.* Western 4.9.59.
1088 N5540   L-188C: *D.* Eastern 17.9.59.
1089 N5007K   *D.* National 4.9.59.
1090 N9705C   *D.* Braniff 18.9.59. *W.O.* 29.9.59.
1091 N171PS   L-188C: *D.* PSA 6.11.59.
1092 N5008K   *D.* National 22.9.59.
1093 N6122A   *D.* American Airlines 23.9.59. *Albany.*
1094 N7139C   *D.* Western 26.9.59.
1095 N9706C   *D.* Braniff 1.10.59.
1096 N5009K   *D.* National 5.10.59.
1097 N5010K   *D.* National 10.10.59.
1098 N5541   L-188C: *D.* Eastern 14.10.59.
1099 N9707C   *D.* Braniff 17.10.59.
1100 N6123A   *D.* American Airlines 22.10.59. *Nashville.*
1101 N125US   L-188C: *D.* Northwest 22.10.59.
1102 N6124A   *D.* American Airlines 23.10.59. *El Paso.*
1103 N111   L-188C: *D.* FAA.
1104 N5011K   *D.* National 13.11.59.
1105 N126US   L-188C: *D.* Northwest 14.11.59.
1106 N9708C   *D.* Braniff 17.11.59.
1107 N5012K   *D.* National.
1108 N127US   L-188C: *D.* Northwest.
1109 N172PS   L-188C: *D.* PSA.
1110 N173PS   L-188C: *D.* PSA.
1111 N128US   L-188C: *D.* Northwest.
1112 N129US   L-188C: *D.* Northwest.
1113 N130US   L-188C: *D.* Northwest.
1114 N9709C   *D.* Braniff.
1115 N6125A   *D.* American Airlines 31.12.59. *Oklahoma City.*
1116 N6126A   *D.* American Airlines 13.1.60. *Tucson.*
1117 N6127A   *D.* American Airlines .1960. *W.O.* 14.9.60.
1118 N7140C   *D.* Western.
1119 N6128A   *D.* American Airlines 27.1.60. *San Diego.*
1120 N6129A   *D.* American Airlines 3.2.60. *Los Angeles.*
1121 N6130A   *D.* American Airlines 12.2.60. *San Francisco.*
1122 N6131A   *D.* American Airlines 17.2.60. *Little Rock.*
1123 N6132A   *D.* American Airlines 19.2.60. *Richmond.*
1124 N6133A   *D.* American Airlines 24.2.60. *Baltimore.*
1125 N6134A   *D.* American Airlines 25.2.60. *Memphis.*
1126 N6135A   *D.* American Airlines 22.3.60. *San Antonio.*
1127 N7141C   *D.* Western.
1128 N7142C   *D.* Western.
1129 N7143C   *D.* Western.
1130 N181H   L-188C: *L.* US Navy Pacific Missile Range. *D.* PSA. 1962.
1131 N131US   L-188C: *D.* Northwest.
1132 N132US   L-188C: *D.* Northwest.
1133 N182H   L-188C: *D.* Sports Aloft 1962.
1134 N183H   L-188C: *D.* Sports Aloft 1962.
1135 N184H   L-188C: *D.* Braniff 5.62.
1136 N185H   L-188C: *D.* Sports Aloft 1962.
1137 N133US   L-188C: *D.* Northwest.
1138 N134US   L-188C: *D.* Northwest.
1139 N135US   L-188C: *D.* Northwest.
1140 N9744C   *D.* Western.
1141 N136US   L-188C: *D.* Northwest.
1142 N138US   L-188C: *D.* Northwest.

1143 N9745C   *D.* Western.
1144 N137US   L-188C: *D.* Northwest. *W.O.* 17.9.61.
1145 N9746C   *D.* Western.
1146 N5013K   *D.* National.
1147 VH-TLC   *D.* TAA.
1148 N5014K   *D.* National.
1149 N6015K   *D.* National.
2001 PH-LLA   L-188C: *F.F.* as N6934C. *D.* KLM 29.9.59. *Mercurius.*
2002 VH-ECA   L-188C: *D.* Qantas 30.10.59. *Pacific Electra.*
2003 PH-LLB   L-188C: *D.* KLM 10.10.59. *Venus.*
2004 VH-ECB   L-188C: *D.* Qantas 30.10.59. *Pacific Explorer.*
2005 ZK-TEA   L-188C: *D.* TEAL 9.11.59. *Aotearoa.*
2006 PH-LLC   L-188C: *D.* KLM. *Mars.*
2007 VH-ECC   L-188C: *D.* Qantas. 25.11.59. *Pacific Endeavour.*
2008 VH-ECD   L-188C: *D.* Qantas. 3.12.59. *Pacific Enterprise.*
2009 PH-LLD   L-188C: *D.* KLM. *Jupiter.*
2010 ZK-TEB   L-188C: *D.* TEAL.
2011 ZK-TEC   L-188C: *D.* TEAL.
2012 PH-LLE   L-188C: *D.* KLM. *Saturnus.*
2013 PH-LLF   L-188C: *D.* KLM. *Uranus.*
2014 PH-LLG   L-188C: *D.* KLM. *Neptunus.*
2015 PH-LLH   L-188C: *D.* KLM. *Pluto.*
2016 PH-LLI   L-188C: *D.* KLM. *Ceres.*
2017 PH-LLK   L-188C: *D.* KLM. *Pallas.*
2018 PH-LLL   L-188C: *D.* KLM. *Orion.*
2019 PH-LLM   L-188C: *D.* KLM. *Sirius. W.O.* 12.6.61
2020 PK-GLA   L-188C: *D.* Garuda. *Palem Bali.*
2021 PK-GLB   L-188C: *D.* Garuda.
2022 PK-GLC   L-188C: *D.* Garuda.

SUD-AVIATION CARAVELLE

01 F-BHHH   Proto.: *F.F.* 27.5.66 as F-WHHH.
02 F-BHHI   Proto.: *F.F.* 24.5.56 as F-WHHI.
 1 F-BHRA   I/111: *F.F.* 18.5.58. *D.* Air France 2.4.59. *Alsace.*
 2 F-BHRB   I/111: *D.* Air France 19.3.59. *Lorraine.*
 3 LN-KLH   I/III: *D.* SAS 10.4.59. *Finn Viking.*
 4 SE-DAA   I/III: *D.* SAS 25.4.59. *Eskil Viking.*
 5 F-BHRC   I/III: *D.* Air France 15.5.59. *Anjou.*
 6 OY-KRA   I/III: *D.* SAS 5.5.59. *Vagn Viking.*
 7 LN-KLI   I/III: *D.* SAS 17.6.59. *Einar Viking.*
 8 F-BHRD   I/III: *D.* Air France 30.6.59. *Guyenne.*
 9 F-BHRE   I/III: *D.* Air France 31.7.59. *Artois.*
10 PP-VJC   I: *D.* Varig 16.9.59. *W.O.* 27.9.61.
11 SE-DAB   I/III: *D.* SAS 18.10.59. *Ingemar Viking.*
12 F-BHRF   I/III: *D.* Air France 17.10.59. *Auvergne.*
13 F-BHRG   I/III: *D.* Air France 6.11.59. *Berry.*
14 OY-KRB   I/III: *D.* SAS 18.11.59. *Orm Viking. W.O.* 19.1.60.
15 PP-VJD   I/III: *D.* Varig 11.12.59.
16 F-BHRH   I/III: *D.* Air France 11.12.59. *Bourgogne.*
17 F-BHRI   I/III: *D.* Air France 18.12.59. *Bretagne.*
18 F-OBNG   I/III: *D.* Air Algérie 26.12.59.
19 LV-   Built as prototype Series III F-WJAQ. *F.F.* 30.12.59. *Cvtd.* to Srs.VI and
       F.F. 10.9.60. *S.* Aerolineas 1.62.
20 F-OBNH   I/III: *D.* Air Algérie 1.1.60. *S.* Varig 1962.
21 OH-LEA   IA/III: *D.* Finnair 18.2.60. *Similintu.*
22 OH-LEB   IA/III: *D.* Finnair 30.3.60. *Sinisiipi.*
23 F-BHRJ   I/III: *D.* Air France 9.2.60. *Champagne.*
24 LN-KLP   IA/III: *D.* SAS 8.3.60. *Troud Viking.*
25 SE-DAC   IA/III: *D.* SAS 11.3.60. *Arne Viking.*
26 F-BHRK   IA/III: *D.* Air France 18.3.60. *Corse.*
27 OH-LEC   IA/III: *D.* Finnair 5.5.60. *Sininuoli.*

28 F-OBNI I/III: *D.* Air Algérie 19.3.60. *S.* Air Liban 1962.
29 OY-KRC IA/III: *D.* SAS 2.4.60. *Faste Viking.*
30 LN-KLR IA/III: *D.* SAS 15.4.60. *Hall Viking.*
31 F-BHRL IA/III: *D.* Air France 26.4.60. *Dauphine.*
32 CN-CCV IA/III: *D.* Royal Air Maroc 11.5.60.
33 HB-ICW IA/III: *D.* Swissair (*L.* from SAS) 30.4.60. *Solothurn.*
34 SE-DAD IA/III: *D.* SAS 28.5.60. *Torolf Viking.*
35 I-DAXA III: *D.* Alitalia 29.4.60. *Altar.*
36 I-DAXE III: *D.* Alitalia 19.5.60. *Aldebaran.*
37 F-BHRM III: *D.* Air France 3.6.60. *Flandre.*
38 HB-ICX III: *D.* Swissair (*L.* from SAS) 24.6.60. *Chur.*
39 F-BHRN III: *D.* Air France. *Gascogne.*
40 I-DAXI III: *D.* Alitalia 10.6.60. *Antares.*
41 F-BHRO III: *D.* Air France 29.6.60. *Ile de France.*
42 F-BJAO III/VII: *D.* General Electric Co. 18.7.60 as N420GE. *Santa Maria. F.F.*
CJ-805-23Cs. 12.60. *Sud demonstrator, Series 10a, 1962.*
43 HB-ICY III: *D.* Swissair (*L.* from SAS) 8.7.60. *Lausanne.*
44 I-DAXO III: *D.* Alitalia 18.7.60. *Deneb.*
45 F-BHRP III: *D.* Air France 12.8.60. *Languedoc.*
46 F-BHRQ III: *D.* Air France 26.8.60. *Limousin.*
47 OY-KRD III: *D.* SAS 30.7.60. *Vif Viking.*
48 HB-ICZ III: *D.* Swissair (*L.* from SAS) 13.8.60. *Bellinzona.*
49 OY-KRE III: *D.* SAS 21.9.60. *Knud Viking.*
50 F-BHRR III: *D.* Air France 23.9.60. *Lyonnais.*
51 F-OBNJ IA/III: *D.* Air Algérie 12.9.60.
52 F-BHRZ III: *D.* Air France 16.3.61. *Flandre.*
53 F-BJTA III: *D.* Air France 7.4.61. *Comte de Nice.*
54 F-BHRS III: *D.* Air France 26.10.60. *Normandie.*
55 F-BHRT III: *D.* Air France 14.10.60. *Picardie.*
56 SE-DAE III: *D.* SAS 28.10.60. *Alric Viking.*
57 CN-CCX III: *D.* Royal Air Maroc 25.2.61.
58 F-BHRU III: *D.* Air France 10.11.60. *Poitou.*
59 F-BHRV III: *D.* Air France 22.11.60. *Provence.*
60 F-BHRX III: *D.* Air France 29.11.60. *Savoie.*
61 F-BHRY III: *D.* Air France 15.12.60. *Touraine.*
62 F-BJAP Proto. VIR: *F.F.* in United colours. 6.2.61.
63 F- XA: Proto. long fuselage. CJ805-23C fans.
64 OO-SRA VI: *D.* Sabena 20.1.61.
65 OO-SRB VI: *D.* Sabena 1.2.61.
66 OO-SRC VI: *D.* Sabena 17.2.61.
67 OO-SRE VI: *D.* Sabena 1.3.61.
68 F-BJTB III: *D.* Air France 30.5.61. *Béarn. W.O.* 12.9.61.
69 OO-SRD VI: *D.* Sabena.
70 OO-SRG VI: *D.* Sabena 12.7.61.
71 I-DABA III: *D.* Alitalia 21.3.61. *Regolo.*
72 I-DABE VIN: *D.* Alitalia 30.3.61. *Rigel.*
73 F-OBNK VI: *D.* Air Algérie 2.5.61.
74 I-DABI VIN: *D.* Alitalia 22.4.61. *Sirio.*
75 F-OBNL VI: *D.* Air Algérie 20.6.61.
76 OO-SRF VI: *S.* Sabena 28.4.61.
77 I-DABU VIN: *D.* Alitalia 17.5.61. *Vega.*
78 OO-SRH VI: *D.* Sabena 11.8.61.
79 I-DAXU III: *D.* Alitalia 9.6.61. *Canopo.*
80 I-DAXT VIN: *D.* Alitalia 17.10.61.
81 I-DABR VIN: *D.* Alitalia.
82 I-DABZ VIN: *D.* Alitalia.
83 OD-ADY III: *D.* Air Liban (*L.*) 10.7.61 (ex Air France F-BJTC).
84 TS-IKM III: *D.* Tunis Air (*L.*) 31.8.61 (ex Air France F-BJTD).
85 I-DABT VIN: *D.* Alitalia 2.4.62.
86 N100IU VIR: *D.* United 10.6.61. *Ville de Toulouse.*
87 N1002U VIR: *D.* United 20.6.61.

88 N1003U VIR: *D.* United 2.2.61.
89 N1004U VIR: *D.* United 6.7.61.
90 N1005U VIR: *D.* United 31.7.61. *Ville de Grenoble.*
91 N1006U VIR: *D.* United 11.8.61. *Ville de Saintes.*
92 N1007U VIR: *D.* United 30.8.61. *Ville de Coutances.*
93 N1008U VIR: *D.* United 20.9.61. *Ville de Rochefort.*
94 N1009U VIR: *D.* United 7.10.61. *Ville de Rouen.*
95 N1010U VIR: *D.* United 20.10.61. *Ville de Strasbourg.*
96 N1011U VIR: *D.* United 26.10.61. *Ville de Dijon.*
97 N1012U VIR: *D.* United 11.11.61. *Ville de Lille.*
98 N1013U VIR: *D.* United 23.11.61. *Ville d'Arles.*
99 N1014U VIR: *D.* United 30.11.61. *Ville de Nice.*
100 N1015U VIR: *D.* United 8.12.61. *Ville de Saint Nazaire.*
101 N1016U VIR: *D.* United 19.12.61. *Ville de Nantes.*
102 N1017U VIR: *D.* United 19.12.61. *Ville de Cannes.*
103 N1018U VIR: *D.* United 4.1.62. *Ville de Bordeaux.*
104 N1019U VIR: *D.* United 6.1.62. *Ville de Lyon.*
105 F-BJTG III: *D.* Air France. 4.5.62. *Roussillon.*
106 I-DABS VIN: *D.* Alitalia. 6.4.62.
107 EC-ARI VIR: *D.* Iberia. 3.62. *Isaac de Albeniz.*
108 EC-ARJ VIR: *D.* Iberia.
109 EC-ARK VIR: *D.* Iberia. 3.62.
110 EC-ARL VIR: *D.* Iberia. 18.4.62.
111 F-BJTC III: *D.* Air France 1.62.
112 HB-ICS III: *D.* Swissair 18.3.62. *Uri.*
113 F-BJTD III: *D.* Air France. 3.62.
114 N1020U VIR: *D.* United 24.2.62. *Ville de Calais.*
115 F-BJTE III: *D.* Air France.
116 OH-LED III: *D.* Finnair. *Sinipiika.*
117 —
118 —
119 F-BJTH III: *D.* Air France.
120 —
121 HB-ICT III: *D.* Swissair. 3.62.
122 HB-ICU III: *D.* Swissair. 3.62.
123 HB-ICV III: *D.* Swissair. 19.4.62.
124 F-BJTF III: *D.* Air France.

VICKERS VANGUARD

703 G-AOYW V.951: *F.F.* 20.1.59 Vickers demonstrator.
704 G-APEA V.951: *F.F.* 22.4.59. *D.* BEA 27.3.61. *Euryalus.*
705 G-APEB V.951: *F.F.* 23.7.59. *D.* BEA 17.3.61. *Bellerophon.*
706 G-APEC V.951: *F.F.* 17.10.59. *D.* BEA 14.1.61. *Sirius.*
707 G-APED V.951: *F.F.* 24.12.59. *D.* BEA 30.1.61. *Vanguard.*
708 G-APEE V.951: *F.F.* 3.2.60. *D.* BEA 14.3.60. *Defiance.*
709 G-APEF V.951: *F.F.* 5.4.60. *D.* BEA 13.12.60. *Victory.*
710 G-APEG V.953: *F.F.* 1.5.61. *D.* BEA 19.5.61. *Arethusa.*
711 G-APEH V.953: *F.F.* 1.6.61. *D.* BEA 21.6.61. *Audacious.*
712 G-APEI V.953: *F.F.* 4.7.61. *D.* BEA 20.7.61. *Indefatigable.*
713 G-APEJ V.953: *F.F.* 28.7.61. *D.* BEA 17.8.61. *Ajax.*
714 G-APEK V.953: *F.F.* 30.8.61. *D.* BEA 16.9.61. *Dreadnought.*
715 G-APEL V.953: *F.F.* 22.9.61. *D.* BEA 7.10.61. *Leander.*
716 G-APEM V.953: *F.F.* 13.10.61. *D.* BEA 3.11.61. *Agamemnon.*
717 G-APEN V.953: *F.F.* 28.10.61. *D.* BEA 14.11.61. *Valiant.*
718 G-APEO V.953: *F.F.* 15.11.61. *D.* BEA 27.11.61. *Orion.*
719 G-APEP V.953: *F.F.* 29.11.61. *D.* BEA 13.12.61. *Superb.*
720 G-APER V.953: *F.F.* 12.12.61. *D.* BEA 1.1.62. *Amethyst.*
721 G-APES V.953: *F.F.* 8.1.62. *D.* BEA 24.1.62. *Swiftsure.*
722 G-APET V.953: *F.F.* 1.2.62. *D.* BEA 21.2.62. *Temeraire.*
723 G-APEU V.953: *F.F.* 8.3.62. *D.* BEA 30.3.62. *Undaunted.*
724 CF-TKA V.952: *F.F.* 21.5.60. *D.* TCA 6.11.61.

VICKERS VANGUARD—*continued*

725 CF-TKB V.952: *F.F.* 25.7.60. *D.* TCA 15.4.61.
726 CF-TKC V.952: *F.F.* 6.10.60. *D.* TCA 14.12.60.
727 CF-TKD V.952: *F.F.* 24.10.60. *D.* TCA 7.12.60.
728 CF-TKE V.952: *F.F.* 5.11.60. *D.* TCA 1.2.61.
729 CF-TKF V.952: *F.F.* 19.11.60. *D.* TCA 7.1.61.
730 CF-TKG V.952: *F.F.* 10.12.60. *D.* TCA 19.1.61.
731 CF-TKH V.952: *F.F.* 15.12.60. *D.* TCA 18.2.61.
732 CF-TKI V.952: *F.F.* 9.1.61. *D.* TCA 11.3.61.
733 CF-TKJ V.952: *F.F.* 19.4.61. *D.* TCA 5.5.61.
734 CF-TKK V.952: *F.F.* 10.5.61. *D.* TCA 24.5.61.
735 CF-TKL V.952: *F.F.* 11.2.61. *D.* TCA 24.3.61.
736 CF-TKM V.952: *F.F.* 15.3.61. *D.* TCA 27.4.61.
737 CF-TKN V.952: *F.F.* 18.4.61. *D.* TCA 1.5.61,
738 CF-TKO V.952: *F.F.* 7.6.61. *D.* TCA 26.6.61.
739 CF-TKP V.952: *F.F.* 30.5.61. *D.* TCA 16.6.61.
740 CF-TKQ V.952: *F.F.* 1.7.61. *D.* TCA 12.8.61.
741 CF-TKR V.952: *F.F.* 17.7.61. *D.* TCA 19.8.61.
742 CF-TKS V.952: *F.F.* 28.8.61. *D.* TCA 28.10.61.
743 CF-TKT V.952: *F.F.* 21.9.61. *D.* TCA 7.10.61.
744 CF-TKX V.952: *F.F.*   . *D.* TCA   .
745 CF-TKY V.952: *F.F.*   . *D.* TCA   .
746 CF-TKZ V.952: *F.F.*   . *D.* TCA   .

VICKERS VISCOUNT

1 G-AHRF V.630: *F.F.* 16.7.48. VX217 for early flights. *W.O.* 27.8.52.
2 VX217 V.663: Ordered as V.609 G-AHRG. *F.F.* (Tay engines) 15.3.50.
3 G-AMAV V.700: *F.F.* 28.8.50. Last flight 4.58. *Sc.* 1961.
4 G-ALWE V.701: *F.F.* 20.8.52. *D.* BEA 3.1.53. *Discovery. W.O.* 14.3.57.
5 G-ALWF V.701: *F.F.* 3.12.52. *D.* BEA 13.2.53. *Sir John Franklin.*
6 G-AMNY V.701: *F.F.* 7.1.53. *D.* BEA 20.2.53. *Sir Ernest Shackleton. W.O.* 5.1.60.
7 G-AMOG V.701: *F.F.* 11.2.53. *D.* BEA 27.3.53. *Robert Falcon Scott.*
8 F-BGNK V.708: *F.F.* 11.3.53. *D.* Air France 18.5.53. *W.O.* 12.12.56.
9 G-AMOA V.701: *F.F.* 27.3.53. *D.* BEA 17.4.53. *Sir George Vancouver.*
10 G-ARBY V.708: *F.F.* 27.5.53. *D.* Air France 25.8.53 as F-BGNL. *D.* Maitland Drewery 6.60. *L.* Danish Air Charter 1961. *L.* BUA 1962.
11 G-AMOB V.701: *F.F.* 10.4.53. *D.* BEA 24.4.53. *William Baffin.*
12 G-ARER V.708: *F.F.* 3.7.53. *D.* Air France 4.6.53 as F-BGNM. *D.* Maitland Drewery 9.60. *L.* BKS Air Transport 29.5.61. *L.* BUA 1962.
13 G-AMOC V.701: *F.F.* 14.5.53. *D.* BEA 4.6.53. *Richard Chancellor.*
14 G-ARGR V.708: *F.F.* 26.8.53. *D.* Air France 29.10.53 as F-BGNN. *D.* Maitland Drewery 6.60. *L.* BKS Air Transport 20.6.61. *L.* BUA 1962.
15 G-AMOD V.701: *F.F.* 19.6.53. *D.* BEA 26.6.53. *John Davis.*
16 F-BGNO V.708: *F.F.* 19.9.53. *D.* Air France 12.12.53. *S.* Air Inter. 1962.
17 G-AMOE V.701: *F.F.* 26.6.53. *D.* BEA 13.7.53. *Sir Edward Parry.*
18 F-BGNP V.708: *F.F.* 13.10.53. *D.* Air France 21.12.53. *S.* Air Inter. 1962.
19 G-AMOF V.701: *F.F.* 23.7.53. *D.* BEA 8.8.53. *Sir Martin Frobisher.*
20 G-AMNZ V.701: *F.F.* 3.9.53. *D.* BEA 3.10.53. *James Cook.*
21 G-AMOH V.701: *F.F.* 1.10.53. *D.* BEA 14.10.53. *Henry Hudson.*
22 G-AMOI V.701: *F.F.* 20.10.53. *D.* BEA 5.11.53. *Sir Hugh Willoughby.*
23 G-AMOJ V.701: *F.F.* 4.11.53. *D.* BEA 30.11.53. *Sir James Ross.*
24 G-AMOK V.701: *F.F.* 21.11.53. *D.* BEA 22.12.53. *Sir Humphrey Gilbert.*
25 G-AMOL V.701: *F.F.* 8.12.53. *D.* BEA 1.1.54. *David Livingstone.*
26 G-AMOM V.701: *F.F.* 22.12.53. *D.* BEA 27.1.54. *James Bruce. W.O.* 21.1.56.
27 G-AMON V.701: *F.F.* 16.1.54. *D.* BEA 11.3.54. *Thomas Cavendish.*
28 G-AMOO V.701: *F.F.* 1.12.53. *D.* BEA 19.12.53. *John Oxenham.*
29 G-AMOP V.701: *F.F.* 8.1.54. *D.* BEA 19.12.53. *Mungo Park.*
30 G-APZB V.707: *F.F.* 30.1.54. *D.* Aer Lingus 7.3.54 as EI-AFV. *St. Patrick. S.* Tradair.
31 G-ARKH V.707: *F.F.* 16.2.54. *D.* Aer Lingus 7.3.54 as EI-AFW. *St. Brigid. S.* Eagle (Bermuda) as VR-BBJ 22.2.60. To Cunard Eagle 5.4.61.

32 VR-BBH V.707: *F.F.* 12.3.54. *D.* Aer Lingus 26.3.54 as EI-AFY. *St. Brendan. S.* Eagle (Bermuda) as VR-BBH 23.1.60. To Cunard Eagle 18.4.61 as G-ARKI. Reverted to VR-BBH in 1962.
33 F-BGNQ V.708: *F.F.* 14.4.54. *D.* Air France 27.5.54. *S.* Air Inter, 1962.
34 G-APZC V.707: *F.F.* 14.3.54. *D.* Aer Lingus 3.4.54 as EI-AGI. *St. Laurence O'Toole. S.* Tradair.
35 F-BGNR V.708: *F.F.* 6.5.54. *D.* Air France 29.6.54. *S.* Air Inter. 1962.
36 G-ARIR V.708: *F.F.* 30.5.54. *D.* Air France 30.6.54 as F-BGNS. Starways 3.2.61.
37 F-BGNT V.708: *F.F.* 16.6.54. *D.* Air France 7.7.54. *L.* Air Vietnam 1961.
38 F-BGNU V.708: *F.F.* 15.7.54. *D.* Air France 28.7.54. *L.* Air Vietnam 1961.
39 F-BGNV V.708: *F.F.* 29.7.54. *D.* Air France 19.8.54. *S.* Air Inter. 1962.
40 CF-TGI V.724: *F.F.* 13.10.54. *D.* TCA 8.12.54.
41 CF-TGJ V.724: *F.F.* 24.12.54. *D.* TCA 30.1.55.
42 CF-TGK V.724: *F.F.* 14.2.55. *D.* TCA 25.2.55.
43 CF-TGL V.724: *F.F.* 13.2.55. *D.* TCA 14.3.55. *W.O.* 10.11.58.
44 VH-TVA V.720: *F.F.* 29.8.54. *D.* TAA 5.10.54. *John Batman. W.O.* 31.10.54.
45 VH-TVB V.720: *F.F.* 20.10.54. *D.* TAA 25.11.54. *Gregory Blackland.*
46 VH-TVC V.720: *F.F.* 17.11.54. *D.* TAA 8.12.54. *John Oxley. L.* Ansett-ANA. *W.O.* 30.11.61.
47 VH-TVD V.720: *F.F.* 1.12.54. *D.* TAA 18.12.54. *Hamilton Hume.*
48 VH-TVE V.720: *F.F.* 23.12.54. *D.* TAA 12.1.55. *Charles Sturt. L.* Ansett-ANA 1961.
49 VH-TVF V.720: *F.F.* 30.1.55. *F.* TAA 2.4.55. *Ernest Giles. L.* Ansett-ANA 1961.
50 CF-TGM V.724: *F.F.* 15.3.55. *D.* TCA 30.3.55.
51 CF-TGN V.724: *F.F.* 31.3.55. *D.* TCA 8.4.55.
52 CF-TGO V.724: *F.F.* 20.4.55. *D.* TCA 1.5.55.
53 CF-TGP V.724: *F.F.* 10.5.55. *D.* TCA 19.5.55.
54 CF-TGQ V.724: *F.F.* 27.5.55. *D.* TCA 5.6.55.
55 CF-TGR V.724: *F.F.* 12.6.55. *D.* TCA 21.6.55.
56 CF-TGS V.724: *F.F.* 2.7.55. *D.* TCA 10.7.55.
57 CF-TGT V.724: *F.F.* 21.7.55. *D.* TCA 4.8.55.
58 CF-TGU V.724: *F.F.* 10.8.55. *D.* TCA 18.8.55.
59 CF-TGV V.724: *F.F.* 31.8.55. *D.* TCA 15.9.55.
60 CF-TGW V.724: *F.F.* 7.2.56. *D.* TCA 15.2.56.
61 G-ANHA V.701: *F.F.* 7.10.54. *D.* BEA 19.10.54. *Anthony Jenkinson.*
62 G-ANHB V.701: *F.F.* 2.11.54. *D.* BEA 21.11.54. *Sir Henry Stanley.*
63 G-ANHC V.701: *F.F.* 1.12.54. *D.* BEA 19.12.54. *Sir Leo McClintock. W.O.* 22.10.58.
64 G-ANHD V.701C: *F.F.* 24.4.55. *D.* BEA 4.5.55. *Sir William Dampier.*
65 G-ANHE V.701C: *F.F.* 22.6.55. *D.* BEA 29.6.55. *Gino Watkins.*
66 G-ANHF V.701C: *F.F.* 4.7.55. *D.* BEA 11.7.55. *Matthew Flinders.*
67 YI-ACK V.735: *F.F.* 23.9.55. *D.* Iraqi A/W. *Ibn Ferras.*
68 YI-ACL V.735: *F.F.* 14.10.55. *D.* Iraqi A/W. *Ibn Sinbad.*
69 YI-ACM V.735: *F.F.* 8.11.55. *D.* Iraqi A/W. *Ibn Bottootah.*
70 CF-GXK V.737: *F.F.* 15.3.55. *D.* Canadian DOT.
71 VP-BBW V.702: *F.F.* 17.6.55. *L.* BWIA as VP-TBK. *D.* Kuwait Airways 26.3.59 as G-APTA. *D.* Bahamas Airways 28.5.61.
72 G-APOW V.702: *F.F.* 6.8.55. *D.* BWIA as VP-TBL. *D.* Kuwait Airways 18.9.58.
73 VP-BBX V.702: *F.F.* 6.9.55. *D.* BWIA as VP-TBM. *D.* Kuwait Airways 16.11.58 as G-APPX. *D.* Bahamas Airways 6.61.
74 G-ANRR V.732: *F.F.* 26.4.55. *D.* Hunting 11.5.55. *L.* MEA as OD-ACF 9.55 to 9.57. *W.O.* 2.12.58.
75 SU-AKY V.732: *F.F.* 7.6.55. *D.* Hunting 15.6.55 as G-ANRS. *L.* MEA as OD-ACH 10.55 to 9.57. *S.* Misrair 31.8.59.
76 SU-AKX V.732: *F.F.* 5.7.55. *D.* Hunting 22.7.55 as G-ANRT. *L.* MEA as OD-ACG 10.55 to 9.57. *L.* Iraqi A/W as YI-ADM 5.58 to 2.59. *S.* Misrair 15.7.59.
77 G-AODG V.736: *F.F.* 20.10.55. *D.* Fred Olsen 18.11.55 as LN-FOF. *L.* BEA as G-AODG. *Fridtjof Nansen.* 15.12.55. *L.* MEA as OD-ACR 27.3.57. *S.* Airwork (now BUA) 1957.
78 G-AODH V.736: *F.F.* 8.11.55. *D.* Fred Olsen 14.12.55 as LN-FOL. *L.* BEA

12.55 as G-AODH. *Roald Amundsen.* L. BWIA as VP-TBY. S. Airwork (now BUA) 1957. *W.O.* 30.10.61.

79	**IU683**	Y.723: *F.F.* 8.11.55. *D.* Indian Air Force 29.12.55.
80	**IU684**	V.730: *F.F.* 14.12.55. *D.* Indian Air Force 13.1.56. *Raj Humsa.*
81	**VP-TBN**	V.702: *F.F.* 5.10.55. *D.* BWIA 4.11.55.
82	**YS-09C**	V.763: *F.F.* 7.10.58. *D.* TACA 31.10.58. *W.O.* 5.3.59.
83	**J751**	V.734: *F.F.* 29.2.56. *D.* Pakistan Air Force 20.3.56.
84	**VH-TVG**	V.720: *F.F.* 8.1.56. *D.* TAA 22.1.56. *William Hovell.*
85	**SU-AIC**	V.739: *F.F.* 1.12.55. *D.* Misrair 23.12.55. *W.O.* 1.10.56.
86	**SU-AID**	V.739: *F.F.* 12.12.55. *D.* Misrair (now UAA) 4.2.45. *W.O.* 16.3.62.
87	**SU-AIE**	V.739: *F.F.* 12.1.56. *D.* Misrair (now UAA) 27.1.56.
88	**G-APKJ**	V.744: *F.F.* 12.5.55. *D.* Capital 15.6.55 as N7402. *S.* Vickers 1958. *L.* All Nippon 30.6.60. *W.O.* 12.6.61.
89	**XR801**	V.744: *F.F.* 30.6.55. *D.* Capital 9.7.55 as N7403. *S.* Vickers 1958 as G-APKK. *L.* All Nippon 23.7.60. *S.* ETPS 1962.
90	**N7404**	V.744: *F.F.* 14.7.55. *D.* Capital 23.7.55. *W.O.* 20.2.56.
91	**CU-T-603**	V.755: Ordered Airwork as G-AOCA. *F.F.* 26.4.56. *D.* Cubana 16.5.56. *W.O.* 1.11.58.
92	**VR-BBL**	V.755: Ordered Airwork as G-AOCB. *F.F.* 24.5.56. *D.* Cubana 5.6.56 as CU-T-604. *S.* Cunard Eagle 1961.
93	**VR-BBM**	V.755: Ordered Airwork as G-AOCC. *F.F.* 16.6.56. *D.* Cubana 20.6.56 as CU-T-605. *S.* Cunard Eagle 1961.
94	**YV-C-AMV**	V.749: *F.F.* 19.1.56. *D.* LAV 28.2.56.
95	**YV-C-AMX**	V.749: *F.F.* 8.2.56. *D.* LAV 13.3.56.
96	**YV-C-AMY**	V.749: *F.F.* 10.2.56. *D.* LAV 23.3.56.
97	**VH-BAT**	V.747: *F.F.* 15.9.55 as G-ANXV. *D.* Butler (now Ansett-ANA) 27.9.55. *Warrel.*
98	**VP-YNA**	V.748: *F.F.* 28.3.56. *D.* CAA 1.5.56. *Malvern.*
99	**VP-YNB**	V.748: *F.F.* 5.5.56. *D.* CAA 17.5.56. *Matopos.*
100	**VP-YNC**	V.748: *F.F.* 24.5.56. *D.* CAA 8.6.56. *Mlanje.*
101	**VP-YND**	V.748: *F.F.* 14.6.56. *D.* CAA 1.7.56. *Mweru.*
102	**VP-YNE**	V.748: *F.F.* 7.7.56. *D.* CAA 18.7.56. *Mpika. W.O.* 9.8.58.
103	**N7405**	V.745: *F.F.* 3.11.55. *D.* Capital (United) 14.11.55.
104	**N7406**	V.745: *F.F.* 19.11.55. *D.* Capital (United) 26.11.55.
105	**N7407**	V.745: *F.F.* 1.12.55. *D.* Capital (United) 7.12.55.
106	**N7408**	V.745: *F.F.* 12.12.55. *D.* Capital (United) 23.12.55.
107	**N7409**	V.745: *F.F.* 20.12.55. *D.* Capital (United) 31.12.55.
108	**N7410**	V.745: *F.F.* 6.1.56. *D.* Capital (United) 15.1.56. *W.O.* 20.5.58.
109	**N7411**	V.745: *F.F.* 20.1.56. *D.* Capital (United) 8.3.56.
110	**N7412**	V.745: *F.F.* 2.2.56. *D.* Capital (United) 23.2.56.
111	**N7413**	V.745: *F.F.* 13.2.56. *D.* Capital (United) 23.2.56.
112	**OE-LAN**	V.745: *F.F.* 22.2.56 as N7414. *D.* Capital (United) 3.3.56. *S.* Austrian A/W 1961. *Johannes Brahms.*
113	**OE-LAD**	V.745: *F.F.* 2.3.56 as N7415. *D.* Capital 13.3.56. *S.* Austrian A/W 1961. *Franz Lehar.*
114	**I-LIRC**	V.745: *F.F.* 10.3.56 as N7416. *D.* Capital 20.3.56. *S.* Alitalia 1961.
115	**N7417**	V.745: *F.F.* 21.3.56. *D.* Capital (United) 29.3.56.
116	**I-LIRE**	V.745: *F.F.* 27.3.56 as N7418. *D.* Capital 30.3.56. *S.* Alitalia 1961.
117	**N7419**	V.745: *F.F.* 10.4.56. *D.* Capital (United) 20.4.56.
118	**PI-C-773**	V.745: *F.F.* 17.4.56 as N7421. *D.* Capital 29.4.56. To Vickers a G-ARHY 1961. *S.* PAL 1961.
119	**I-LITZ**	V.745: *F.F.* 24.4.56 as N7421. *D.* Capital 2.5.56. *S.* Alitalia 1961.
120	**N7422**	V.745: *F.F.* 1.5.56. *D.* Capital (United) 7.5.56.
121	**N7423**	V.745: *F.F.* 8.5.56. *D.* Capital (United) 14.5.56.
122	**N7424**	V.745: *F.F.* 15.5.56. *D* Capital (United) 19.5.56.
123	**N7425**	V.745: *F.F.* 23.5.56. *D.* Capital (United) 29.5.56.
124	**N7426**	V.745: *F.F.* 29.5.56. *D.* Capital (United) 31.5.56.
125	**N7427**	V.745: *F.F.* 5.6.56. *D.* Capital (United) 12.6.56.
126	**N7428**	V.745: *F.F.* 12.6.56. *D.* Capital (United) 20.6.56.
127	**N7429**	V.745: *F.F.* 19.6.56. *D.* Capital (United) 30.6.56.
128	**N7430**	V.745: *F.F.* 22.6.56. *D.* Capital (United) 2.7.56.
129	**N7431**	V.745: *F.F.* 6.7.56. *D.* Capital (United) 17.7.56.
130	**I-LIFS**	V.745: *F.F.* 16.7.56 as N7432. *D.* Capital 20.7.56. *S.* Alitalia 1961.
131	**I-LINS**	V.745: *F.F.* 22.7.56 as N7433. *D.* Capital 26.7.56. *S.* Alitalia 1961.
132	**N7434**	V.745: *F.F.* 30.7.56. *D.* Capital (United) 3.8.56.
133	**N7435**	V.745: *F.F.* 10.8.56. *D.* Capital (United) 18.8.56.
134	**N7436**	V.745: *F.F.* 12.8.56. *D.* Capital (United) 17.8.56.
135	**N7437**	V.745: *F.F.* 24.8.56. *D.* Capital 31.8.56. *W.O.* 6.4.58.
136	**N7438**	V.745: *F.F.* 1.9.56. *D.* Capital (United) 6.9.56.
137	**N7439**	V.745: *F.F.* 8.9.56. *D.* Capital (United) 13.9.56.
138	**N7440**	V.745: *F.F.* 14.9.56. *D.* Capital (United) 24.9.56.
139	**N7441**	V.745: *F.F.* 17.9.56. *D.* Capital (United) 23.9.56.
140	**TF-ISN**	V.759: *F.F.* 2.11.56 as G-AOGG. *D.* Hunting 16.11.56. *S.* Icelandair 4.57. *Gullfaxi.*
141	**FAB-2100**	V.742: *F.F.* 24.7.56 as LN-SUN for Braathens. *S.* Brazil AF. *D.* 1.2.57.
142	**CF-TGX**	V.757: *F.F.* 18.3.56. *D.* TCA 28.3.56.
143	**CF-TGY**	V.757: *F.F.* 22.3.56. *D.* TCA 29.3.56. *W.O.* 3.10.59.
144	**CF-TGZ**	V.757: *F.F.* 14.4.56. *D.* TCA 12.5.56.
145	**VH-BUT**	V.747: *F.F.* 27.8.56 as G-ANYH. *D.* Butler (now Ansett-ANA) 14.9.56. *Warrawee II.*
146	**VH-TVH**	V.756: *F.F.* 3.5.56. *D.* TAA 3.6.56. *George Bass.*
147	**VH-TVI**	V.756: *F.F.* 26.5.56. *D.* TAA 9.6.56. *Matthew Flinders.*
148	**VH-TVJ**	V.756: *F.F.* 14.6.56. *D.* TAA 25.6.56. *John Forrest.*
149	**TS-IST**	V.759: *F.F.* 7.12.56. as G-AOGH. *D.* Hunting 28.12.56. *S.* Icelandair 4.57. *Hrinfaxi.*
150	**G-AOJA**	V.802: *F.F.* 27.7.56. *D.* BEA 14.2.57. *Sir Samuel White Baker. W.O.* 23.10.57.
151	**G-AOJB**	V.802: *F.F.* 29.9.56. *D.* BEA 6.2.57. *Stephen Borough.*
152	**G-AOJC**	V.802: *F.F.* 7.11.56. *D.* BEA 19.1.57. *Robert O'Hara Burke.*
153	**G-AOJD**	V.802: *F.F.* 29.11.56. *D.* BEA 11.1.57. *Sebastian Cabot.*
154	**G-AOJE**	V.802: *F.F.* 1.1.57. *D.* BEA 26.1.57. *Sir Alexander Mackenzie.*
155	**G-AOJF**	V.802: *F.F.* 18.1.57. *D.* BEA 8.2.57. *Sir George Somers.*
156	**G-AOHG**	V.802: *F.F.* 4.2.57. *D.* BEA 20.2.57. *Richard Hakluyt.*
157	**G-AOHH**	V.802: *F.F.* 15.2.57. *D.* BEA 5.3.57. *Sir Robert McClure.*
158	**G-AOHI**	V.802: *F.F.* 26.2.57. *D.* BEA 11.3.57. *Charles Montagu Doughty.*
159	**G-AOHJ**	V.802: *F.F.* 7.3.57. *D.* BEA 24.3.57. *Sir John Manderville.*
160	**G-AOHK**	V.802: *F.F.* 20.3.57. *D.* BEA 5.4.57. *John Hanning Speke.*
161	**G-AOHL**	V.802: *F.F.* 29.3.57. *D.* BEA 17.4.57. *Charles Sturt.*
162	**G-AOHM**	V.802: *F.F.* 30.5.57. *D.* BEA 27.6.57. *Robert Machin.*
163	**G-APHN**	V.802: *F.F.* 13.4.57. *D.* BEA 1.5.57. *Alexander Gordon Laing.*
164	**G-AOHO**	V.802: *F.F.* 36.4.57. *D.* BFA 4.5.57. *Samuel Wallis.*
165	**G-AOHP**	V.802: *F.F.* 4.5.57. *D.* BEA 17.5.57. *James Weddel. W.O.* 17.11.57.
166	**G-AOHR**	V.802: *F.F.* 22.5.57. *D.* BEA 4.6.57. *Sir Richard Burton.*
167	**G-AOHS**	V.802: *F.F.* 7.6.57. *D.* BEA 22.6.57. *Robert Thorne.*
168	**G-AOHT**	V.802: *F.F.* 20.6.57. *D.* BEA 3.7.57. *Ralph Fitch.*
169	**G-AOHU**	V.802: *F.F.* 28.6.57. *D.* BEA 11.7.57. *Sir George Strong Naves. W.O.* 7.1.60.
170	**G-AOHV**	V.802: *F.F.* 10.7.57. *D.* BEA 25.7.57. *Sir John Barrow.*
171	**G-AORD**	V.802: *F.F.* 22.8.57. *D.* BEA 7.9.57. *Arthur Philipp.*
172	**PH-VIA**	V.803: *F.F.* 5.4.57. *D.* KLM 6.6.57. *Sir Sefton Brancker.*
173	**PH-VIB**	V.803: *F.F.* 24.6.57. *D.* KLM 26.7.57. *Louis Blériot.*
174	**PH-VIC**	V.803: *F.F.* 19.7.57. *D.* KLM 29.8.57. *J. C. Ellehammer.*
175	**PH-VID**	V.803: *F.F.* 12.8.56. *D.* KLM 17.8.56. *Otto Lilienthal.*
176	**PH-VIE**	V.803: *F.F.* 11.9.57. *D.* KLM 19.10.57. *Jan Olieslagers.*
177	**PH-VIF**	V.803: *F.F.* 28.9.56. *D.* KLM 22.10.57. *Leonardo da Vinci.*
178	**PH-VIG**	V.803: *F.F.* 18.10.57. *D.* KLM 8.11.57. *Sir Charles Kingsford Smith.*
179	**PH-VIH**	V.803: *F.F.* 1.11.57. *D.* KLM 23.11.57. *Wright Bros.*
180	**PH-VII**	V.803: *F.F.* 22.11.57. *D.* KLM 23.12.57. *Daidalos.*
181	**VH-TVK**	V.756: *F.F.* 10.1.57. *D.* TAA 18.1.57. *Thomas Mitchell.*
182	**G-AOFX**	V.801C: *F.F.* 16.7.56. *D.* BEA 22.7.56. *Sir Joseph Banks.*
183	**N905**	V.764: *F.F.* 2.10.56. *D.* U.S. Steel Corp. 4.12.56.
184	**N906**	V.764: *F.F.* 26.10.56. *D.* U.S. Steel Corp. 29.11.56.

185   N907   V.764: *F.F.* 30.11.56. *D.* U.S. Steel Corp. 28.12.56.
186   9M-ALY   V.760: *F.F.* 13.12.56 as VR-HFI. *D.* Hong Kong Airways 1.1.57. *S.* Malayan Airways.
187   VR-SEE   V.760: *F.F.* 14.2.57 as VR-HFJ. *D.* Hong Kong Airways 28.2.57. *S.* Malayan Airways.
188   XY-ADF   V.761: *F.F.* 26.6.57. *D.* UBA.
189   XY-ADG   V.761: *F.F.* 27.8.57. *D.* UBA.
190   XY-ADH   V.761: *F.F.* 26.9.57 as XY-ADH. *D.* UBA. *L.* Kuwait Airlines as G-APZN 1960.
191   N306   V.765: *F.F.* 24.1.57. *D.* Standard Oil Co. 11.2.57.
192   VT-DIO   V.768: *F.F.* 30.7.57. *D.* Indian Airlines 19.8.57.
193   VT-DIF   V.768: *F.F.* 11.9.57. *D.* Indian Airlines 26.9.57.
194   VT-DIG   V.768: *F.F.* 10.10.57. *D.* Indian Airlines 23.10.57.
195   VT-DIH   V.768: *F.F.* 30.10.57. *D.* Indian Airlines 9.11.57.
196   VT-DII   V.768: *F.F.* 6.11.57. *D.* Indian Airlines 21.11.57.
197   VH-TVL   V.756: *F.F.* 22.2.57. *D.* TAA 1.3.57. James Cook.
198   XR802   V.745: *F.F.* 24.9.56 as N7442. *D.* Capital 29.9.56. *L.* Northeast. To Vickers as G-ARUU 20.6.61. *S.* ETPS 1962.
199   N7443   V.745: *F.F.* 2.10.56. *D.* Capital (United) 5.10.56.
200   N7444   V.745: *F.F.* 7.10.56. *D.* Capital (United) 13.10.56.
201   N7445   V.745: *F.F.* 11.10.56. *D.* Capital (United) 17.10.56.
202   N7446   V.745: *F.F.* 17.10.56. *D.* Capital (United) 23.10.56.
203   N7447   V.745: *F.F.* 22.10.56. *D.* Capital (United) 29.10.56.
204   N7448   V.745: *F.F.* 28.10.56. *D.* Capital (United) 3.11.56.
205   N7449   V.745: *F.F.* 31.10.56. *D.* Capital (United) 7.11.56.
206   N7450   V.745: *F.F.* 9.11.56. *D.* Capital (United) 16.11.56.
207   N7451   V.745: *F.F.* 14.11.56. *D.* Capital (United) 23.11.56.
208   N7452   V.745: *F.F.* 19.11.56. *D.* Capital (United) 30.11.56.
209   N7454   V.745: *F.F.* 26.11.56. *D.* Capital (United) 11.12.56.
210   N7455   V.745: *F.F.* 1.12.56. *D.* Capital (United) 16.12.56.
211   N7456   V.745: *F.F.* 6.12.56. *D.* Capital (United) 13.12.56.
212   N7457   V.745: *F.F.* 15.12.56. *D.* Capital (United) 31.12.56.
213   N7458   V.745: *F.F.* 27.12.56. *D.* Capital (United) 6.1.57.
214   N7459   V.745: *F.F.* 28.12.56. *D.* Capital (United) 17.1.57.
215   N7460   V.745: *F.F.* 3.1.57. *D.* Capital (United) 9.1.57.
216   N7461   V.745: *F.F.* 24.1.57. *D.* Capital (United) 1.2.57.
217   N7462   V.745: *F.F.* 2.2.57. *D.* Capital 2.3.57. *W.O.* 18.1.60.
218   CF-THA   V.757: *F.F.* 14.1.57. *D.* TCA 5.2.57.
219   CF-THB   V.757: *F.F.* 18.1.57. *D.* TCA 29.1.57.
220   CF-THC   V.757: *F.F.* 11.2.57. *D.* TCA 20.2.57.
221   CF-THD   V.757: *F.F.* 16.2.57. *D.* TCA 22.2.57.
222   CF-THE   V.757: *F.F.* 24.2.57. *D.* TCA 5.3.57.
223   CF-THF   V.757: *F.F.* 10.3.57. *D.* TCA 14.3.57.
224   CF-THG   V.757: *F.F.* 19.3.57. *D.* TCA 28.3.57.
225   G-APNF   V.776: *F.F.* 26.2.57 as G-16-4. *L.* Aer Lingus 3.4.58 as EI-AJW. *L.* BEA 1958 as G-APNF. *Philipp Carteret. S.* Kuwait Oil Co. 21.10.58.
226   N6599C   V.798: *F.F.* 22.3.57 as V.745. Vickers demonstrator G-APBH 5.57 *D.* Northeast Airlines 27.2.59.
227   PI-C-772   V.784B: *F.F.* 10.4.57 as V.745. *L.* TAA as VH-TVO *David Lindsay.* 14.3.58–31.1.59. *D.* PAL 17.12.59.
228   CF-RBC   V.793D: *F.F.* 23.4.57 as V.745, G-16-3. *L.* Aer Lingus as EI-AJV 27.3.58. *L.* BEA as G-APNG *James Lancaster* 1958. *L.* Kuwait Airways 13.9.58. *D.* Royal Bank of Canada 21.4.59. *S.* Canadian Breweries. *S.* S. J. Grooves & Sons.
229   CF-DTA   V.745: *F.F.* 28.4.57. Used by Vickers as G-APFR. *D.* Canadian DOT 28.10.58.
230   N6595C   V.798: *F.F.* 7.6.57 as V.745 G-16-6. Used by Vickers as G-APLX. *S.* Northeast Airlines 2.10.58.
231   N7465   V.745: *F.F.* 18.6.57. *D.* Capital, 19.6.58.
232   N6590C   V.798: *F.F.* 28.6.57. *D.* Northeast 8.5.58.

233   N6591C   V.798: *F.F.* 27.6.57. *D.* Northeast 8.8.58.
234   N6592C   V.798: *F.F.* 5.7.57. *D.* Northeast 21.8.58.
235   VP-TBS   V.772: *F.F.* 6.10.57. *D.* BWIA 26.11.57.
236   VP-TBT   V.772: *F.F.* 30.10.57. *D.* BWIA 13.11.57.
237   VP-TCU   V.772: *F.F.* 19.11.57. *D.* BWIA 11.12.57.
238   VP-TBX   V.772: *F.F.* 8.12.57. *D.* BWIA 3.2.58.
239   OD-ACT   V.754: *F.F.* 17.7.57. *D.* MEA 29.7.57.
240   JY-ACI   V.754: *F.F.* 29.8.57 as OD-ACU 9.9.57. *S.* Jordanian 1961.
241   VP-YTE   V.754: *F.F.* 19.9.57 as OD-ACV. *D.* MEA 27.9.57. *S.* CAA 1961.
242   OD-ACW   V.754: *F.F.* 2.10.57. *D.* MEA 9.10.57.
243   JY-ACK   V.754: *F.F.* 22.10.57 as OD-ADD. *D.* MEA 9.11.57. *S.* Jordanian 1961.
244   OD-ADE   V.754: *F.F.* 24.11.57. *D.* MEA 12.12.57.
245   OD-ACX   V.754: *F.F.* 21.12.57. *D.* MEA 4.1.58.
246   TC-SEG   V.794: *F.F.* 9.1.58 as OD-ACY. *D.* THY 2.58.
247   VT-DOD   V.779: *F.F.* 11.3.57 as LN-FOM. *D.* Olsen 30.4.57. *L.* Austrian Airlines as OE-LAE 1958–60. *L.* BEA as G-ARBW. *L.* SAS. *S.* Indian Airlines 1962.
248   G-AOXU   V.804: *F.F.* 31.8.57. *D.* Transair (now BUA) 17.9.57.
249   G-AOXV   V.804: *F.F.* 18.9.57. *D.* Transair (now BUA) 24.9.57.
250   VT-DOE   V.779: *F.F.* 13.4.57 as LN-FOH. *D.* Olsen 25.4.57. *L.* Austrian Airlines as OE-LAB. *L.* BEA as G-APZP. *S.* Indian Airlines 1962.
251   VT-DOH   V.779: *F.F.* 7.5.57 as LN-FOI. *D.* Olsen 28.5.57. *L.* Austrian Airlines as OE-LAC. *L.* SAS. *S.* Indian Airlines 1962.
252   VT-DOI   V.779: *F.F.* 22.7.57 as LN-FOK. *D.* Olsen 9.8.57. *L.* Austrian Airlines as OE-LAD. *S.* Indian Airlines 1962.
253   G-AOHW   V.802: *F.F.* 18.7.57. *D.* BEA 1.8.57. *Sir Francis Younghusband.*
254   G-AORC   V.802: *F.F.* 29.7.57. *D.* BEA 17.8.57. *Richard Lauder. W.O.* 28.4.58.
255   G-AOYF   V.806A: *F.F.* 9.8.57. Vickers prototype. *W.O.* 20.10.57.
256   G-AOYG   V.806: *F.F.* 4.10.57. *D.* BEA 29.3.58. *Charles Darwin.*
257   G-AOYI   V.806: *F.F.* 14.11.57. *D.* BEA 1.1.58. *Sir Humphry Davy.*
258   EI-AMA   V.805: *F.F.* 2.12.57 as G-APDW. *D.* Eagle 19.12.57. To Eagle (Bermuda) 3.58 as VR-BAX. *Enterprise. S.* MCA as CF-MCJ 1959. *S.* Aer Lingus 1961. *Connlaodh.*
259   G-AOYS   V.806: *F.F.* 9.12.57. *D.* BEA 8.1.58. *Edward Jenner.*
260   G-AOYK   V.806: *F.F.* 13.1.58. *D.* BEA 12.2.58. *Edmund Cartwright.*
261   G-AOYL   V.806: *F.F.* 23.1.58. *D.* BEA 3.2.58. *Lord Joseph Lister.*
262   G-AOYM   V.806: *F.F.* 26.2.58. *D.* BEA 19.3.58. *John Loudon McAdam.*
263   G-AOYN   V.806: *F.F.* 6.3.58. *D.* BEA 26.3.58. *Sir Isaac Newton.*
264   G-AOYO   V.806: *F.F.* 18.3.58. *D.* BEA 3.4.58. *Adam Smith.*
265   G-AOYP   V.806: *F.F.* 21.3.58. *D.* BEA 16.5.58. *John Napier.*
266   G-AOYR   V.806: *F.F.* 21.3.58. *D.* BEA 11.4.58. *Sir Richard Arkwright.*
267   G-AOYS   V.806: *F.F.* 24.5.58. *D.* BEA 13.6.58. *George Stephenson.*
268   G-AOYT   V.806: *F.F.* 21.4.58. *D.* BEA 2.5.58. *James Watt.*
269   CF-THH   V.757: *F.F.* 12.5.57. *D.* TCA 19.5.57.
270   CF-THI   V.757: *F.F.* 19.5.57. *D.* TCA 26.5.57.
271   CF-THK   V.757: *F.F.* 28.5.57. *D.* TCA 3.6.57.
272   CF-THL   V.757: *F.F.* 29.11.57. *D.* TCA 16.12.57.
273   CF-THM   V.757: *F.F.* 8.12.57. *D.* TCA 18.12.57.
274   CF-THN   V.757: *F.F.* 18.12.57. *D.* TCA 3.1.58.
275   CF-THO   V.757: *F.F.* 18.1.58. *D.* TCA 24.1.58.
276   CF-THP   V.757: *F.F.* 22.1.58. *D.* TCA 29.1.58.
277   CF-THQ   V.757: *F.F.* 2.2.58. *D.* TCA 14.2.58.
278   CF-THR   V.757: *F.F.* 17.2.58. *D.* TCA 27.2.58.
279   CF-THS   V.757: *F.F.* 27.1.58. *D.* TCA 1.2.58.
280   150   V.781: *F.F.* 18.5.58. *D.* SAAF 16.6.58.
281   ZK-BRD   V.807: *F.F.* 11.12.57. *D.* NZNAC 31.12.57. *City of Wellington.*
282   ZK-BRE   V.807: *F.F.* 17.1.59. *D.* NZNAC 31.1.59. *City of Auckland.*
283   ZK-BRF   V.807: *F.F.* 24.2.59. *D.* NZNAC 12.3.59. *City of Christchurch.*
284   N6594C   V.798: *F.F.* 31.5.57 as V.745. *D.* Northeast 22.9.58.
285   N7464   V.745: *F.F.* 21.8.57. *D.* Capital 20.2.58.
286   N6593C   V.798: *F.F.* 19.8.58 as V.745. *D.* Northeast 3.9.58.

287 **N7463** V.745: *F.F.* 8.12.57. *D.* Capital 25.1.58. *W.O.* 12.5.59.
288 **N6596C** V.798: *F.F.* 21.10.58 as V.745. *D.* Northeast 31.10.58.
289 **EI-AJI** V.808: *F.F.* 13.5.57. *D.* Aer Lingus 22.5.57. *Gall.*
290 **EI-AJJ** V.808: *F.F.* 24.12.57. *D.* Aer Lingus 10.1.58. *Columban.*
291 **EI-AJK** V.808: *F.F.* 13.2.58. *D.* Aer Lingus 21.2.58. *Cillian.*
292 **VT-DIX** V.768: *F.F.* 17.2.58. *D.* IAC 3.3.58.
293 **VT-DIZ** V.768: *F.F.* 6.3.58. *D.* IAC 18.3.58.
294 **VT-DJA** V.768: *F.F.* 8.4.58. *D.* IAC 21.4.58.
295 **VT-DJB** V.768: *F.F.* 17.5.58. *D.* IAC 22.5.58.
296 **VT-DJC** V.768: *F.F.* 30.5.58. *D.* IAC 6.6.58.
297 **EP-AHA** V.782: *F.F.* 4.3.58. *D.* Iranian Airlines 25.3.58.
298 **EP-AHB** V.782: *F.F.* 29.3.58. *D.* Iranian Airlines 16.4.58.
299 **EP-AHC** V.782: *F.F.* 30.4.58. *D.* Iranian Airlines 9.5.58.
300 **PI-C-770** V.784: *F.F.* 1.5.57. *D.* PAL 10.5.57.
301 **CF-THJ** V.757: *F.F.* 23.5.57. *D.* TCA 30.5.57.
302 **CF-THT** V.757: *F.F.* 23.2.58. *D.* TCA 1.3.58.
303 **CF-THU** V.757: *F.F.* 3.3.58. *D.* TCA 8.3.58.
304 **CF-THV** V.757: *F.F.* 13.3.58. *D.* TCA 21.3.58.
305 **CF-THW** V.757: *F.F.* 25.3.58. *D.* TCA 3.4.58.
306 **CF-THX** V.757: *F.F.* 13.4.58. *D.* TCA 18.4.58.
307 **CF-THY** V.757: *F.F.* 23.4.58. *D.* TCA 5.4.58.
308 **CF-THZ** V.757: *F.F.* 10.5.58. *D.* TCA 15.5.58.
309 **CF-TIA** V.757: *F.F.* 7.6.58. *D.* TCA 16.6.58.
310 **CF-TIB** V.757: *F.F.* 5.6.58. *D.* TCA 11.6.58.
311 **G-AOYH** V.806: *F.F.* 25.10.57. *D.* BEA 23.12.58. *William Harvey.*
312 **EI-ALG** V.805: *F.F.* 3.2.58. *D.* Eagle Airways 5.3.58 as G-APDX. To Eagle (Bermuda) 19.2.59 as VR-RAY. *Good Fortune. S.* Aer Lingus 2.60.
316 **PP-SRH** V.810: *F.F.* 23.12.57 as G-AOYV. Vickers prototype. *S.* VASP 10.60.
317 **ZS-CVA** V.818: *F.F.* 3.11.58 as CU-T-621. *D.* Cubana 4.8.59. *S.* SAA 3.62.
318 **ZS-CVB** V.818: *F.F.* 19.11.58 as CU-T-622. *D.* Cubana 4.8.59. *S.* SAA 3.62.
319 **VH-RML** V.818: *F.F.* 25.6.59 as CU-T-623. *D.* Cubana 20.8.59. *S.* Ansett-ANA 3.62.
320 **N500T** V.818: *F.F.* 4.9.59 as CR-T-624. *D.* Cubana. *S.* Tennessee Gas Transmission. *D.* 24.9.59.
321 **CX-AQN** V.769: *F.F.* 20.1.58. *D.* PLUNA 13.5.58.
322 **CX-AQO** V.769: *F.F.* 3.2.58. *D.* PLUNA 13.5.58.
323 **CX-AQP** V.769: *F.F.* 15.2.58. *D.* PLUNA 11.2.60.
324 **PI-C-771** V.784: *F.F.* 5.9.57. *D.* PAL 18.9.57. *L.* TACA as YS-06C 1957–1958.
325 **I-LIFE** V.785: *F.F.* 12.3.57. *D.* LAI (Alitalia) 27.3.57.
326 **I-LIFT** V.785: *F.F.* 11.4.57. *D.* LAI (Alitalia) 17.4.57.
327 **I-LILI** V.785: *F.F.* 18.5.57. *D.* LAI (Alitalia) 26.5.57.
328 **I-LAKE** V.785: *F.F.* 18.8.57. *D.* LAI (Alitalia) 30.8.57.
329 **I-LARK** V.785: *F.F.* 23.6.57. *D.* LAI (Alitalia) 30.6.57.
330 **I-LOTT** V.785: *F.F.* 19.7.57. *D.* LAI (Alitalia) 28.7.57.
331 **YI-ACU** V.773: *F.F.* 30.10.57. *D.* Iraqi Airways 18.11.58. *Ibn Turaik.*
332 **YS-08C** V.786: *F.F.* 5.7.57 as HK-943X (Lloyd Aereo Colombiano). *D.* 23.8.57. *Santa Margarita. S.* TACA 1960.
333 **YS-011C** V.786: *F.F.* 23.12.57 as HK-946X (Lloyd Aereo Colombiano). *S.* Lancia as AN-AKP. *Nicarao.* 23.1.58. *S.* TACA 1960.
334 **N200Q** V.786: *F.F.* 19.1.58 as HK-947X (Lloyd Aereo Colombiano). *S.* LANCIA as AN-AKQ *Ruben Dario.* 27.2.58. *S. H. H.* May, 1960.
335 **AP-AJC** V.815: *F.F.* 17.12.58. *D.* PIA 12.1.59. *City of Karachi. W.O.* 18.5.59.
336 **AP-AJD** V.815: *F.F.* 27.1.58. *D.* PIA 13.2.59. *City of Lahore.*
337 **AP-AJE** V.815: *F.F.* 15.2.59. *D.* PIA 17.3.59. *City of Dacca. W.O.* 14.8.59.
338 **D-ANUN** V.814: *F.F.* 22.9.58 *D.* Lufthansa 5.10.58.
339 **D-ANOL** V.814: *F.F.* 1.12.58. *D.* Lufthansa 19.12.58.
340 **D-ANAD** V.814: *F.F.* 15.12.58. *D.* Lufthansa 10.1.59.
341 **D-ANIP** V.814: *F.F.* 31.1.59. *D.* Lufthansa 15.2.59. *L.* Condor Flugdienst 2.11.61.
342 **D-ANUR** V.814: *F.F.* 16.2.59. *D.* Lufthansa 8.3.59.
343 **D-ANEF** V.814: *F.F.* 10.3.59. *D.* Lufthansa 26.3.59.
344 **D-ANIZ** V.814: *F.F.* 20.3.59. *D.* Lufthansa 6.4.59.

345 **FAB-2101** V.789: *F.F.* 1.12.57. *D.* Brazil Govt. 6.10.58.
346 **ZS-CDT** V.813: *F.F.* 17.9.58. *D.* SAA 26.10.58. *Blesbok.*
347 **ZS-CDU** V.813: *F.F.* 14.10.58. *D.* SAA 26.10.58. *Bosbok.*
348 **ZS-CDV** V.813: *F.F.* 7.11.58. *D.* SAA 18.11.58. *Waterbok.*
349 **ZS-CDW** V.813: *F.F.* 26.11.58. *D.* SAA 8.12.58. *Roosbok.*
350 **ZS-CDX** V.813: *F.F.* 7.12.58. *D.* SAA 20.12.58. *Wildebees.*
351 **ZS-CDY** V.813: *F.F.* 22.12.58. *D.* SAA 10.1.59. *Gemsbok.*
352 **ZS-CDZ** V.813: *F.F.* 15.1.59. *D.* SAA 29.1.59. *Hartbees.*
353 **N240V** V.812: *F.F.* 14.2.58. *D.* Continental 10.5.58. *S.* Tennessee Gas 1961.
354 **N243V** V.812: *F.F.* 31.5.58. *D.* Continental 13.6.58.
355 **VH-RMK** V.812: *F.F.* 1.4.58 as N241V. *D.* Continental 10.5.58. *S.* Ansett-ANA 9.60.
356 **N242V** V.812: *F.F.* 6.5.58. *D.* Continental 23.5.58.
357 **N244V** V.812: *F.F.* 19.6.58. *D,* Continental 26.6.58.
358 **N245V** V.812: *F.F.* 1.7.58. *D.* Continental 8.7.58.
359 **N246V** V.812: *F.F.* 12.7.58. *D.* Continental 1.8.58.
360 **N248V** V.812: *F.F.* 20.8.58. *D.* Continental 16.9.58.
361 **N249V** V.812: *F.F.* 9.9.58. *D.* Continental 2.10.58.
362 **N250V** V.812: *F.F.* 1.10.58. (Temporarily G-APPC). *D.* Continental 24.10.58.
363 **N251V** V.812: *F.F.* 13.10.58. *D.* Continental 5.11.58.
364 **N252V** V.812: *F.F.* 25.10.58. (Temporarily G-APPU). *D.* Continental 23.3.59.
365 **N253V** V.812: *F.F.* 6.12.58. *D.* Continental 23.3.59.
366 **N254V** V.812: *F.F.* 2.4.59. *D.* Continental 10.4.59.
368 **D-ANAM** V.814: *F.F.* 8.4.59. *D.* Lufthansa 17.4.59.
369 **D-ANAB** V.814: *F.F.* 16.4.59. *D.* Lufthansa 29.4.59.
370 **D-ANAC** V.814: *F.F.*   . *D.* Flugdienst (Lufthansa) 30.7.61.
371 **9G-AAV** V.838: *F.F.* 6.9.61. *D.* Ghana A/W. 19.6.61.
372 **9G-AAW** V.838: *F.F.* 7.10.61. *D.* Ghana A/W. 21.10.61.
373 **VH-TVM** V.756: *F.F.* 14.6.58. *D.* TAA 4.7.58. *John Fawkner.*
374 **VH-TVN** V.756: *F.F.* 2.7.58. *D.* TAA 8.7.58. *Wlliam Dampier.*
375 **AP-AJF** V.815: *F.F.* 14.8.59. *D.* PIA 31.8.59.
376 **AP-AJG** V.815: *F.F.* 18.9.59. *D.* PIA 28.9.59.
377 **I-LIRS** V.785: *F.F.* 11.3.58. *D.* Alitalia 3.4.58.
378 **I-LIZT** V.785: *F.F.* 30.3.58. *D.* Alitalia 5.4.58. *W.O.* 21.12.59.
379 **I-LIRP** V.785: *F.F.* 23.4.58. *D.* Alitalia 29.4.58.
380 **I-LIZO** V.785: *F.F.* 9.5.58. *D.* Alitalia 15.5.58.
381 **G-APEX** V.806: *F.F.* 13.6.58. *D.* BEA 24.6.58. *John Harrison.*
382 **G-APEY** V.806: *F.F.* 7.7.58. *D.* BEA 18.7.58. *William Murdoch.*
383 **CF-TIC** V.757: *F.F.* 23.6.58. *D.* TCA 30.6.58.
384 **CF-TID** V.757: *F.F.* 25.2.59. *D.* TCA 8.3.59.
385 **CF-TIE** V.757: *F.F.* 10.3.59. *D.* TCA 20.3.59.
386 **CF-TIF** V.757: *F.F.* 23.3.59. *D.* TCA 28.3.59.
387 **CF-TIG** V.757: *F.F.* 16.4.59. *D.* TCA 2.5.59.
389 **N247V** V.812: *F.F.* 23.7.58. *D.* Continental 20.8.58.
391 **N6597C** V.798: *F.F.* 30.11.58. *D.* Northeast 11.12.58.
392 **N6598C** V.798: *F.F.* 21.12.58. *D.* Northeast 8.1.59.
393 **SU-AKN** V.739A: *F.F.* 3.7.58. *D.* Missair (UAA) 12.7.58.
394 **SU-AKO** V.739A: *F.F.* 15.7.58. *D.* Missair 27.7.58.
395 **G-APKG** V.804: *F.F.* 15.4.58. *D* Transair (BUA) 24.4.58.
396 **G-APKF** V.806: *F.F.* 2.7.58. *D.* BEA 12.7.58. *Michael Faraday.*
397 **PP-SRC** V.827: *F.F.* 30.9.58. *D.* VASP 28.10.58.
398 **PP-SRD** V.827: *F.F.* 23.10.58. *D.* VASP 10.11.58.
399 **PP-SRE** V.827: *F.F.* 13.11.58. *D.* VASP 3.12.58.
400 **PP-SRF** V.827: *F.F.* 6.12.58. *D.* VASP 22.12.58.
401 **PP-SRG** V.827: *F.F.* 20.1.59. *D.* VASP 29.1.59. *W.O.* 22.12.59.
402 **G-APND** V.831: *F.F.* 4.2.59. *D.* Airwork (BUA) 23.2.59.
403 **G-APNE** V.831: *F.F.* 9.3.59. *D.* Airwork (BUA) 17.3.59.
412 **G-APIM** V.806: *F.F.* 4.6.58. *D.* BEA 23.6.58. *Robert Boyle.*
413 **G-APJU** V.806: *F.F.* 21.7.58. *D.* BEA 2.8.58. *Sir Gilbert Blane.*
414 **VH-RMG** V.832: *F.F.* 23.2.59. *D.* Ansett-ANA 12.3.59.
415 **VH-RMH** V.832: *F.F.* 19.3.59. *D.* Ansett-ANA 1.4.59.

416	VH-RMI	V.832: *F.F.* 8.4.59. *D.* Ansett-ANA 24.4.59.
417	VH-RMJ	V.832: *F.F.* 6.5.69. *D.* Ansett-ANA 15.5.59.
418	G-APOX	V.806: *F.F.* 23.3.59. *D.* BEA 11.4.59. *Isambard Brunel.*
419	ST-AAN	V.831: *F.F.* 30.5.59. *D.* Sudan A/W. 5.6.59.
421	EI-AKO	V.808: *F.F.* 21.12.58 as EI-AKJ. *D.* Aer Lingus 7.1.59. *Colman.*
422	EI-AKK	V.808: *F.F.* 19.1.59. *D.* Aer Lingus 31.1.59. *Aidan.*
423	EI-AKL	V.808: *F.F.* 11.3.59. *D.* Aer Lingus 20.3.59. *Colmcille.*
424	G-APTB	V.833: *F.F.* 13.5.59. *D.* Hunting Clan (BUA) 18.6.59.
425	G-APTC	V.833: *F.F.* 9.6.59. *D.* Hunting Clan 29.6.59.
426	G-APTD	V.833: *F.F.* 1.7.59. *D.* Hunting Clan 6.2.60.
427	SU-AKW	V.739B: 3.60. *D.* Misrair (UAA) 6.4.60. *W.O.* 29.9.60.
428	ZK-BWO	V.807: *D.* NZNAC 20.5.61. *City of Dunedin.*
429	TC-SEV	V.794: *F.F.* 16.7.58. *D.* THY 1.8.58. *W.O.* 17.2.59.
430	TC-SEL	V.794: *F.F.* 1.9.58. *D.* THY 19.9.58.
431	TC-SES	V.794: *F.F.* 12.10.58. *D.* THY 24.10.58.
432	TC-SET	V.794: *F.F.* 8.11.58. *D.* THY 2.12.58.
433	VH-TVP	V.816: *F.F.* 8.5.59. *D.* TAA 23.5.59. *John Gould.*
434	VH-TVQ	V.816: *F.F.* 8.6.59. *D.* TAA 17.6.59. *McDouall Stuart.*
435	N40N	V.836: *F.F.* 24.8.59 as VH-TVR. *D.* Union Carbide.
436	EP-MRS	V.816: *F.F.* as VH-TVS. *D.* Iran Govt. 15.5.61.
437	OE-LAF	V.837: *F.F.* 10.2.60. *D.* Austrian Airlines 23.2.60. *W.O.* 26.9.60.
438	OE-LAG	V.837: *F.F.* 17.2.60. *D.* Austrian Airlines 29.2.60. *Franz Schubert.*
439	OE-LAH	V.837: *F.F.* 25.2.60. *D.* Austrian Airlines 14.3.60. *Anton Bruckner.*
440	OE-LAK	V.837: *F.F.* 4.3.60. *D.* Austrian Airlines 21.3.60. *Johann Strauss.*
441	OE-LAL	V.837: *F.F.* *D.* Austrian Airlines 14.5.60. *W. A. Mozart.*
442	OE-LAM	V.837: *F.F.* *D.* Austrian Airlines 3.8.60. *Ludwig V. Beethoven.*
443	JA-8201	V.828: *F.F.* as G-ARKX. *D.* All Nippon 21.7.61.
444	JA-8202	V.828: *F.F.* 7.6.61 as G-ARKY. *D.* All Nippon 22.6.61.
445	JA-8203	V.828: *F.F.* 24.8.61 as G-ARKZ. *D.* All Nippon 2.9.61.
446	9G-AAU	V.838: *F.F.* 5.11.61. *D.* Ghana A/W. 26.11.61.
447	D-ANAF	V.818: *F.F.* 9.12.61. *D.* Lufthansa 5.1.62.
448	JA-8204	V.828: *F.F.* as G-ARWT.
449	JA-8205	V.828: *F.F.* as G-ARWU.
450	JA-8206	V.828: *F.F.* as G-ARWV.

VISCOUNT 800 *(continued from p. 113)*

Most of the second series of Viscounts were ordered in this guise, although a few airlines purchased the V800 Series. These have customer designations between 802 and 809, with the exception that the V806 is a "hybrid" for BEA, with R.Da.7 Dart 520 engines to increase the cruising speed but without the full structural modifications of the V810. Another one-customer variant was Hunting Clan's V833, with Dart 530s providing higher powers for "high and hot" airfields.

When the V810 was announced, Vickers stated the airframe was suitable for use with the 2,350-h.p. Dart R.Da.10 Mk. 541 engine. Any aircraft converted to have this power-plant would be designated in the V840 Series and would have a 400-m.p.h. cruising speed.

Further development of the Viscount series through the V850 and V870 project studies led eventually to the Vanguard.

A list of customer type numbers for the V800 and V810 Series Viscounts follows:

V802. Twenty-two for BEA.

V803. Nine for KLM.

V804. Three for Transair (later BUA), sold to LOT, 1962.

V805. Two ordered by Fred Olsen: delivered to Eagle Airways (later Cunard Eagle) for Bermuda company. One sold later to Aer Lingus; the other sold to Maritime Central Airways and then to Aer Lingus.

V806. One Vickers prototype and nineteen for BEA with Dart 520.

V807. Four for NZNAC.

V808. Six for Aer Lingus.

V810. One Vickers prototype of new Series, later sold to VASP.

V812. Fifteen for Continental. One later sold to Ansett-ANA and one to Tennessee Gas Transmission.

V813. Seven for SAA.

V814. Eleven for Lufthansa, including two assigned to subsidiary Condor-Flugdienst.

V815. Five for Pakistan International.

V816. Two for TAA. One for Shah of Persia.

V818. Four ordered for Cubana; three delivered and one sold to Tennessee Gas Transmission. One other sold later to Ansett-ANA and two to SAA.

V827. Five for VASP.

V828. Nine for All Nippon Airways.

V831. Two for Airwork (BUA) and one for Sudan Airways.

V832. Four for Ansett-ANA.

V833. Three for Hunting Clan (BUA).

V835. One (ex V818) for Tennessee Gas Transmission.

V836. One for Union Carbide Co.

V837. Six for Austrian Airlines.

V838. Six for Ghana Airways.

V843. Six for China.